THE
HAND
BOOK

by

Stephanie Brown

ERGONOME
New York

Preventing Computer Injury:
The HAND Book

Published by
ERGONOME INCORPORATED
145 West 96th Street
New York, NY 10025

(212) 222-9600 telephone
(212) 222-6699 fax

Copyright ©1992-93
Stephanie Brown

This book is intended to help prevent Repetitive Stress Injuries, such as Carpal Tunnel Syndrome, acquired from typing at computer keyboards. It is not represented or warranted to provide a cure for existing injury, nor should it be used as a substitute for medical advice; please be sure to consult your physician if you suspect that you may have a Repetitive Stress Injury. Due to the fact that all individuals interpret and use information differently, neither the author nor the publisher can be held responsible or liable to any person or entity with respect to any consequences caused or alleged to be caused, directly or indirectly, by the use of this book.

Library of Congress Catalog Card Number: 93-73771

ISBN 1-884388-01-9

Printed in the United States of America

10 9 8 7 6 5 4 3

TABLE OF CONTENTS

PREFACE

I decided to write this book when I learned of the rapid increase in the number of people who suffer from computer-related hand injuries. Known as Repetitive Stress Injuries (RSI) or Cumulative Trauma Disorders (CTD), these conditions include tendonitis, carpal tunnel syndrome, cysts and bursitis. Symptoms can range from feelings of numbness, tingling, burning and throbbing to weakness and even paralysis in the fingers, hands and arms. Afflicted people face possible surgery, extensive loss of time at work, and, in some cases, even eventual unemployment. Often the simplest of tasks, such as picking up a fork, will be excruciatingly painful for them.

A three-year study by the National Institute for Occupational Safety and Health estimated that more than 20 percent of people whose work is primarily at a computer keyboard are affected.

As a pianist, I have been aware of such injuries since the mid-70's when I was a student at The Juilliard School of Music. Approximately one-quarter of the piano students there suffered from the very same disorders. My roommate, for example, would have red streaks running up and down her arms after playing for an hour, and she was in so much pain that it was often difficult for her to brush her teeth.

The motions used in playing the piano are almost identical to those used in typing on a computer keyboard; a pianist will use more weight, and obviously needs to cover a wider keyboard, but the finger and arm movements are basically the same. Indeed, before the use of computers became widespread, these problems were usually thought of as "pianists' problems."

Although about 25 percent of those piano students experienced pain, there were many who played all day without feeling the slightest strain. Not everyone who uses a computer suffers either. Why can some people type all day with no difficulty, while others will feel acute pain? Is there a crucial difference between the way these two groups of people are using their keyboards? Are some people using positions and motions that virtually guarantee they will injure themselves? The answer to the last two questions is, unfortunately, yes.

The idea that certain motions or positions can cause injury is known in virtually every field of sports. If you play golf or tennis, or swim, chances are you've had lessons to learn the "right" way to perform a stroke. You've worked on it and you've learned how to avoid positions and motions which are dangerous and are known to cause painful conditions like tennis-elbow, bursitis or tendonitis. You have also at one time or another had the experience of watching someone swing a tennis racquet or a golf-club, and thinking to yourself, "Ouch!" It just *looks* wrong.

Typing at a computer keyboard may not be a sport, but with RSI's becoming so prevalent, it is clear that it must be seen as an athletic activity, with many of the risks and dangers of any athletic activity. Typing may be *micro*-athletic, but it is athletic nonetheless. The actual motions used in typing are small and so are the muscles involved, but that just makes the muscles all the more prone to injury. A shoulder or a thigh muscle can take a great deal more use (and abuse) than can a tiny tendon in your hand. The slightest swelling of these tendons, or in the sheaths which protect them, can lead to debilitating pain and make it virtually impossible for one to type, or do much else.

Just as we would take lessons to "fix" something in our golf or tennis game, we need to learn a better way to type if we're experiencing pain, or even slight strain, at the keyboard. The connection has to be made between harmful positions and motions at the keyboard and consequent injury. Many of us learned little more than a-s-d-f-g and semi-l-k-j-h when we studied typing, if we studied it at all. It is not enough. We need to learn how to work at our computers safely.

Until now there have been several solutions for sufferers of RSI's:

Firstly, inventors have been rushing to design new keyboards, some of them requiring that the user learn a totally new typing system. Although others of the new keyboards use the traditional key layout, they are expensive, and while some of them may be beneficial, this book will demonstrate that they are not necessary for safe typing.

For the already injured there is no substitute for the counsel and care of skilled physicians, and the medical profession has developed extensive means of helping such people to manage their injuries. These include physical therapy, anti-inflammatory drugs, anesthetics, muscle relaxants, cryotherapy (cold treatment) and thermotherapy (heat treatment).

Splints can be placed on the hands and forearms of some victims to prevent them from getting into unnatural positions while they're typing. Unfortunately, however, these are effective on only a few of the dangerous positions, and they can sometimes even be harmful because they lock the forearms and wrists into stationary positions. The hand surgeon Peter A. Nathan was quoted in CTD News, an occupational health newsletter, as having said, "The benefits of immobilization through splints and braces is an old wives' tale. Splinting can cause a weakness in forearm muscles that bend and straighten the wrist and fingers."

The constriction of natural motion can be as dangerous to the arms as the unnatural positions which the splints are intended to prevent. Splints are a temporary solution, at best, and when they're off the wearers often go back to the habits that brought them to grief in the first place.

Other remedies include the injection of cortisone to reduce swelling and, in cases of carpal tunnel syndrome, surgery to cut the ligament at the base of the wrist. Obviously, neither is without its unwanted consequences, and even after such serious treatment patients who have not learned safer ways to work at their computers will revert to old habits, with the expectable result. As the Mayo Clinic reported in its Proceedings of July, 1989, "If followed by a return to the same traumatic environment, the operation is often unsuccessful in controlling symptoms."

Still another solution is to find a new occupation — one which doesn't involve working at a computer. Although sometimes companies can offer their employees alternative jobs, obviously this is simply not a possibility for most people. It also leads to another extremely unfortunate phenomenon, namely that *some people are afraid to report these problems for fear of being thought difficult or even getting fired.* Here the afflicted are caught in a double bind, because by putting off treatment they only insure that their problems will grow still worse.

The oldest cure of all is, of course, rest, and sometimes doctors are compelled to advise people that they face a lifetime of pain and perhaps permanent crippling if they don't take off from work for anywhere from a few months to several years. Unfortunately, however, the economic repercussions of this prescription are usually at least as severe as trying to find a new line of work.

As extreme as these solutions are, they do not attack the problem at its origin — namely, what people are doing at their computers to hurt themselves. They are all responses to the *effects* of Repetitive Stress Injuries. The solution proposed in this book is to attack the causes. These problems can be drastically reduced by showing people how to work at their computers safely. RSI begins at the keyboard and must be solved at the keyboard.

The training is quick and very simple — in fact, its success is due in part to its ease of comprehension and the speed with which it can be put into practice. It is my experience that it is effective both in preventing initial injury and in teaching the already injured how to avoid re-injury after medical treatment.

The HAND Book consists of two sections: Part One, "At Your Keyboard", and Part Two, "Caring for Your Hands".

In Part One, you will learn to identify the common positions and motions you may be using at your keyboard which are dangerous for your hands. The illustrations and photographs will help you to recognize the dangerous positions and to learn what the correct and safe positions are. A few simple exercises, all based on your body's natural alignment, will teach you how to *feel* the difference between the harmful positions and the safe ones. Once you've felt this difference, you'll see how much more comfortable your hands are in their most natural (and safe) positions. Many of these adjustments involve a difference of an inch or two, or less. When you're dealing with instruments as delicate as your hands and fingers, though, even a tiny adjustment can mean the difference between health and injury.

If you're thinking that it will take you years to fix your problems at your computer, you're wrong. An important tenet of this book is that when you're shown a more comfortable way to perform an activity, your body will, in effect, soon learn to "choose" that way. Just as you are able to make a quick change when you've been shown a more comfortable way to execute a motion in sports, so will you be able to make these adjustments at your computer easily. All of the exercises can be done in a few minutes. After a day or two, at most a few weeks, you won't even need to remind yourself of them. Your hands will automatically find their natural, safe position.

In Part Two you'll learn techniques such as massage, heat and ice therapy and stretching to help you keep your hands healthy. People who would not dream of going jogging or working out without warming up and stretching think nothing of sitting down at their computers and starting to type with cold hands and wrists. In Chapter Seventeen I'll show you how to properly warm up your hands.

Although many articles on this subject have stressed the importance of taking frequent rest periods, in Chapter Eighteen, "Taking a Break", you'll learn what you can do during these breaks to really *help* your hands. Since finger and hand health depend in part on neck and back health, this chapter also contains good exercises for your back and neck which can help relieve the stiffness that can result from poor posture. Because working at a computer involves long hours of sitting, we will also discuss the right way to sit at your keyboard. A well-designed, adjustable workstation is necessary for safe keyboard work, and in Appendix A you will learn how to precisely adjust your workstation to best suit your body.

Hands and fingers vary greatly from person to person. Each of us has a completely unique hand and hand-shape — as unique as our finger-prints. You might think, then, that it would be impossible for one method to work for everyone, but this is not so. We all have one "perfect" hand position, and that is the one our hand is in when it's relaxed. You will learn how to find your perfect hand position, and how to keep it when you're at your keyboard.

Computing doesn't have to hurt. You can learn how to work safely and comfortably.

Stephanie Brown
New York

Part One
At Your Keyboard

Chapter One

THE NATURAL LINE
AND THE DANGEROUS ANGLE

When we do something in a particular way for a long time, we get used to it, even if it's wrong. That's why bad habits can be hard to break. Your muscles signal to your brain, "Yes, this is right. This is familiar." But what is familiar may not always be right.

Many people sit down to work at their computers and start by putting their hands in this position:

Illustration No. 1

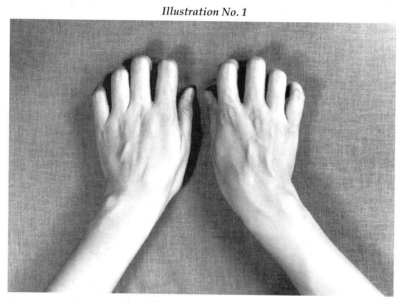

The Dangerous Angle - No

This is one of the biggest mistakes you can make. Although aligning your fingers with the keys in this way may look and feel right to you (and may even be the way you were taught), by doing this you are actually unaligning your hands with your arms. You are creating a dangerous angle in your wrist. Your hands may seem to be in line with your keyboard but they are actually out of line with your body.

Always stay lined up with yourself. Learn to feel when you're doing something at your keyboard that's uncomfortable or even painful for your hands, and learn how to adjust it.

There is an easy way to show you how unnatural this wrist angle is.

Just rest one of your hands on a table, fingertips down, like this:

Illustration No. 2

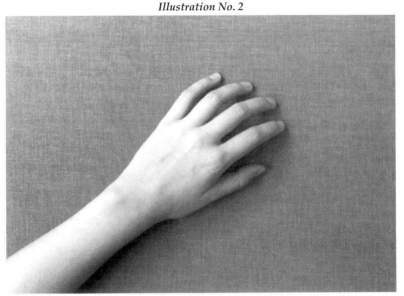

The Natural Line - Yes

Notice the line that goes from your elbow through your forearm and wrist through to your third finger. It's perfectly straight. This is your natural position. This is the way your hands should always look. Remember, any angle in your wrist, if held for more than a moment, is unnatural.

It's just common sense to realize that if you hold any part of your body in an unnatural way for very long it's going to hurt, and if you let it hurt for too long, it will become damaged.

Now, with your hand in this natural position on a table, wiggle your fingers. Notice how it feels — easy and natural, like nothing.

9

Keep wiggling your fingers and slowly angle your wrist until it is in the position of Illustration No. 3 (think of duck feet!). You'll notice that the further you angle it, the worse it feels. You will begin to feel the strain in your wrist and, if you hold it for a minute or two, all the way up in the middle of your forearm.

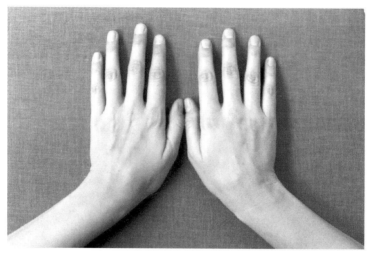

Illustration No. 3

Then, still wiggling your fingers, slowly bring your wrist back until it's in a natural line with your forearm. Your wrist will *naturally* find the position where everything feels easy and comfortable again.

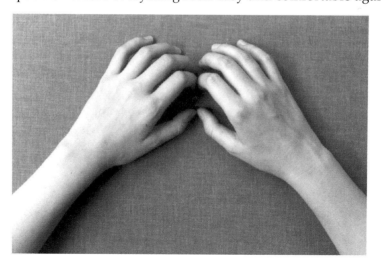

When you put your hands and wrists in the position of Illustration No. 1, literally cramping them so that your fingers will point straight forward, you create that dangerous angle in your wrist, and you may be keeping it that way for many hours every day.

Once you've felt how uncomfortable it is to wiggle your fingers in the angled wrist position, just imagine that many people (and maybe you) are spending anywhere from 20 to 60 hours a week with their wrists in that position. No wonder it hurts! The angled wrist position is unnatural for your body and is one of the leading causes of crippling computer injuries.

EXERCISE No. 1
"The Fluid Line"

Observe yourself as you're walking around, brushing your teeth, turning off a light switch or picking something up. Notice your wrists. Whatever you're doing, they're usually in a straight and *fluid* line with your forearm. Reach for something with your hand and your elbow will follow naturally.

Now, *try* to angle your wrist, as in Illustration No. 1, and do the same things. It's difficult and uncomfortable, and the feeling of fluidity is lost. You have to concentrate very hard to keep your hand angled for even five minutes. Get used to going back and forth between these two positions — angled and non-angled — so that you can really teach your body the difference. Finally, just let your wrists go. You'll feel a sensation of relief and relaxation.

Do this five minutes a day for a day or two, noticing the way these two positions feel. Then go to your keyboard and get into your usual typing position, without actually typing (that you'll learn in the next chapter). Notice if you're been typing with angled wrists. If you have, this position will now look wrong to you. It *should* look wrong. After you've done this exercise it will *feel* wrong as well.

It's easy to teach your body to unlearn a bad habit once you feel how uncomfortable it really is and once you've become aware of an easier and better way. It's a simple thing we're after. You want your hands to look and feel the same at your keyboard as they do everywhere else.

Do the Fluid Line often and remind yourself and your wrists what their natural position really is.

Your body will convince itself.

Chapter Two

DO YOU NEED A NEW KEYBOARD?

After reading the first chapter, you may be thinking to yourself, "I'm going to have to learn how to type all over again." Not true.

You may have seen on the news or read in the paper of the new computer keyboards that are being designed. Their primary objective is to avoid the Dangerous Angle. Most of them look something like this:

Illustration No. 5

Doctors, and now inventors, have become alert to the hazards of working with the wrist in this angled position. An angled keyboard can make it easier to maintain an un-angled wrist, but the simple fact is that you can avoid the angle without getting a new keyboard. Here's how:

The typing position you probably use involves a starting position on the "Home" row with your left hand fingers on A, S, D and F, and your right hand fingers on semi-colon, L, K and J. The thumbs are on the space bar. From this position, you move to accommodate all the other characters. You won't need to change this.

Keep your fingers over the same keys, but just un-angle your wrist as you learned in Chapter One, and let your thumbs separate slightly.

Illustration No. 6

From this ...

Angled-wrist position - No

Illustration No. 7

... to this:

The Natural Line - Yes

Now that you have the basic position, this exercise will help you integrate it into your typing:

EXERCISE No. 2
"Natural Typing"

Find five or ten minutes at your keyboard when you are not under pressure. Place your hands on the keyboard, just as we did in Chapter One, and without typing, move from the natural line (Illustration No. 7) to the angled wrist position (Illustration No. 6), and then back again. Get used to the way these two positions feel at your keyboard — the ease and airy feeling of correct alignment, and the strained feeling of unalignment. Now put your fingers on the keys, and, with your arms and wrists in the natural line, start typing at a slow and gentle pace.

At first it may feel a little different. As you can see, we're talking about a small adjustment — a couple of inches or less. But even if you read no further in this book, these few inches will make a world of difference to your hands.

Stay focused on the fluid line from your elbow through your wrist to your third finger. If you're worried that you might lose some accuracy, relax. You won't. In fact, you'll soon notice that, because your wrists are more relaxed in this position, in time your accuracy will actually *increase*.

For a few weeks, you may notice that your wrists occasionally want to move back to their old positions. That's not unusual — after all, you may have been typing that way for years. If that starts happening, just be patient with yourself. Frustration will only make you tighter.

Give your wrists time to adjust and enjoy their new position.

 TIP: If you're wondering how to reach keys like Return and Tab without angling your wrist, see Chapter Thirteen.

Chapter Three

SITTING WELL

It is well known that sitting carelessly is bad for your back. It is less well known that it can be equally harmful to your wrists and hands. Let's start with the small of your back. If your chair doesn't provide the proper support so that you can easily maintain good posture, you may find yourself compensating after a few hours ...

... by slumping forward like this: ... or by over-arching like this:

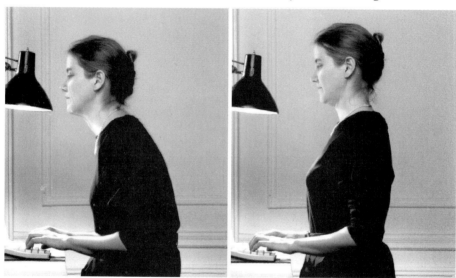

Illustration No. 8   *Illustration No. 9*

In both cases you have forced both your back and your neck into unnatural, cramped and strained positions, and you will feel a lot of discomfort. Find a chair which you feel *fully* supports your back.

It will then have a natural, slight curve and will stay more relaxed. Your neck will not be forced to tilt either forward, if you are slumping, or backward if you are overarching. When your neck is tight, your shoulders also tend to tighten and raise up. Keeping your neck relaxed will help you keep your shoulders down and comfortable. Adjust the height and tilt of your screen so that your neck won't feel any strain when you look at it.

Get up from your chair as often as you can — at least once an hour, more if possible. Stand and stretch, even if it's only for a minute, and it will do you a lot of good. Also, see Chapter Eighteen ("Taking a Break") for some good stretching and strengthening exercises for your back and neck.

After you've found a chair that supports the small of your back, and you've adjusted your screen so that your neck feels comfortable when you look at it, you need to do one more simple thing: adjust the height of your chair. The optimum height is one where your forearms are approximately level and parallel to the floor, with little or no bend in your wrists. Your thighs should also be more or less parallel to the floor. This height allows your shoulders to be relaxed, your elbows loose and your wrists supple.

When your chair is too high, your arms will be at a downward angle to the keyboard, and you'll tend to lean slightly forward as in Illustration No. 10.

Illustration No. 10   *Illustration No. 11*

This position, however, is not as harmful for your hands and wrists as the position you're forced into when your chair is too low, or when you're resting your wrists on a wrist pad or your desk, as in Illustration No. 11.

Sitting too low in relation to your keyboard will cause you to bend and strain your wrists.

No

In this position, you are forcing your wrists to hold up not only themselves and your fingers but the whole weight of your arm as well. Your arm is heavy. Hold the weight of one with the other and you'll see. Your wrists are not meant to be supporting that kind of weight for any length of time. You are also then creating another dangerous wrist position. This one I call the Cobra Position — bending your hand up at the wrist.

This is just as harmful as the Dangerous Angle we discussed in Chapter One. To try it out for yourself, just bend your hands up at the wrists and hold them that way for a minute or two. You'll start getting very tired underneath your wrists. Try to wiggle your fingers in the Cobra Position. It won't feel good. How can you prevent this?

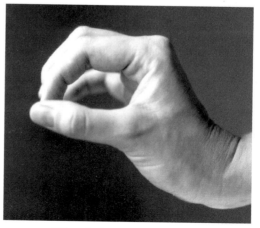

The Cobra Position - No

If your arms are parallel to the floor, there is no need for your wrists to angle up. Also, the device that enables you to increase the tilt of the back of your keyboard must be used cautiously. Too high a tilt will *create* the Cobra Position. You want to maintain a gentle, fluid and approximately level line:

Yes

Find the position that feels comfortable to you — if you prefer, one with a *very slight* bend in your wrist is also safe. You don't ever want to feel that you're "holding up" your arms with your wrists. You don't want to feel even the slightest strain in your wrists. They don't want to be contorted *either* way: angled, as in Chapter One, or bent up as in the illustrations on the opposite page. Your guide will always be how it feels. Angled and bent positions simply don't *feel* good. Notice when you're getting yourself into one, and learn to keep your wrists completely free and relaxed.

EXERCISE No. 3
"The Floating Wrist"

Sit somewhere and drop your hands down at your sides. Slowly angle your wrists up, then let them slowly drop and relax, so they're just hanging from your arms. Do this a few times. Learn to distinguish the difference between the strained feeling of holding your wrists up and the easy feeling of just letting them hang from your shoulders. ⇨

Go to your keyboard, with your chair at its usual height, and see if your wrists feel as if they're holding up your arms. Notice if there is any upward "bend" at your wrists. If there is, you'll want either to raise your chair or lower your keyboard until you feel that your wrists can be completely relaxed. Sometimes it's a small adjustment, but it can be a crucial one.

Imagine floating wrists — completely tension-free.

Typing should be easy and painless.

TIP: Wrist pads are great while you're resting your hands, or reading over work, but not while you're typing. When you type while resting your wrists on a wrist pad or on your desk, once again you're isolating your finger and hand muscles from your arm muscles, just as in the Dangerous Angle and the Cobra Position. Injury comes most often from overusing small muscles and underusing large ones.

TIP 2: If your chair has armrests, use them for resting — not while you're typing.

TIP 3: A well-designed, adjustable workstation is essential for safe keyboard work — for more exact instructions on how to adjust your workstation to fit your body, see APPENDIX A.

Chapter Four

UNGLUING YOUR ELBOWS

Once you've adjusted the height of your chair, you're ready for the other chair adjustment you'll want to make — how far you're sitting from your keyboard. If you sit too close, your elbows will press in so that they become glued to your sides and your wrists will be forced into the Dangerous Angle.

All of the circulation which should be flowing freely throughout your arms — from your shoulders, down through your elbows to your fingers — will be constricted. It will, in effect, be "pinched off" and blood will not be getting through to your hands and fingers where it's needed.

The solution? Clasp your hands and place them on a table, with your elbows free. Notice that you're creating a triangle with your two arms and the front of your body. Your elbows feel loose and suspended, as if they're hanging from your shoulders. Your wrists, without your having to think about it, are perfectly aligned with your arms.

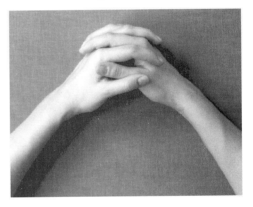

Thinking "triangle" is very helpful. It encourages you to think of your arms as "growing out of" your body, rather than as appendages simply stuck onto your shoulders. Your body functions much more effectively when its various parts work together as an organic whole. Your fingers "borrow" energy and strength from your arms, rather than having to generate them all on their own.

Sit down at your computer. Make sure that the distance from your chair to the keyboard is such that when your hands are in typing position you can see and feel a comfortable and relaxed triangle position. Your elbows will be hanging freely from your shoulders and probably will be lightly touching your sides, although this depends on your particular build.

Once you've found the right distance, you can do the following exercise to help keep your elbows free:

EXERCISE No. 4
"The Chicken-Wing"

Glued elbows feel as bad as they look. Whenever you feel yours getting stuck, stop typing for a minute.

With your hands still at your keyboard, move your elbows in and out, letting your wrists follow along, like a chicken moving its wings. Then slowly let the motion grow smaller and smaller, like a pendulum, until your elbows come to rest.

Where they stop is their perfect position. Your hands and wrists will be in a natural, fluid line. Your elbows may or may not be touching your sides lightly, depending on your physique.

Illustration No. 16

Thinking of the triangle will help you to maintain a continuous flow from your elbows through to your fingertips. Remember, holding any part of your body in a static position will always produce tension.

Do the Chicken-Wing often at your keyboard. It will help you keep your elbows loose and unglued.

Chapter Five

THAWING YOUR WRISTS

Now that you've unglued your elbows, we can move on to relaxing your wrists. Even if you do everything we've suggested so far, and have improved some of your bad positions, you may still have a condition I call Frozen Wrists.

While you're typing, if you feel any rigidity, tightness or tension in your wrists or slightly higher up your arms, then you have Frozen Wrists. When your wrists are not relaxed, all the weight, energy and strength from your arms gets bottled-up in them and can't pass through to your fingers.

Your wrists are very delicate. Once you've learned your natural wrist positions in Chapters One and Three, you'll want to learn how to keep them always relaxed, with the energy from your arms moving through them to your fingertips and out to the keyboard. For this the following exercise can be very helpful:

EXERCISE No. 5
"The Swing"

Rest your right hand on a table, fingertips down, with your palm just above the table. Let a little weight sink into your fingers from your arm until you feel your fingers grow slightly warm. Now take hold of your right elbow with your left hand and gently swing your right elbow forward an inch or two, letting your wrist rise, and then allowing it to fall back down again. Repeat it a few times ... forward and back, just like a swing. Then do it with just your right hand alone.

Keep on doing it, forward and back, and start wiggling your fingers. Find the exact spot in The Swing where wiggling is easiest, and this will be your right position — the precise spot where typing will be most comfortable for you.

Once you've found this position, remember, *don't hold on to it.* Let it be a fluid position. Whenever you feel your wrists tightening up, place your hand on a table and do The Swing. Even just thinking of The Swing can help. The beauty of this exercise is that you simply can't do it with a stiff, frozen wrist.

Give your wrists a break — let them be relaxed.

Chapter Six

THE SPIDER POSITION

The very name of this position is intended to "shock" your hand out of an all too common and very dangerous position. If your fingers look anything like this when you type, you are typing in the Spider Position:

My first computer teacher suffered from an advanced case of the Spider Position. In this position your first knuckles are collapsed, causing your fingers to curl upward and away from your hand. In effect, you've cut off your fingers from the rest of your hand, creating a lot of tension under your wrist and in your knuckles. Hold your fingers in the Spider Position and you can feel it all the way up into your forearm.

No

Your hands should look like this when you're typing:

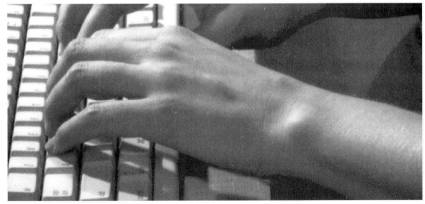

Yes

The Cobra and the Spider are close relatives. The Cobra cuts off your hand from your arm and the Spider cuts off your fingers from your hand. They both constrict vital circulation which should run from your shoulders all the way through your arms, to your fingers. They are completely contrary to your body's natural muscular balance.

There's an easy "antidote" to the Spider Position:

EXERCISE No. 6
"The Rainbow"

Hold out your right hand and get into the Spider Position.

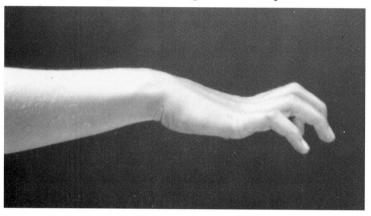

With your left hand, reach under your knuckles and slowly raise the bridge of your right hand.

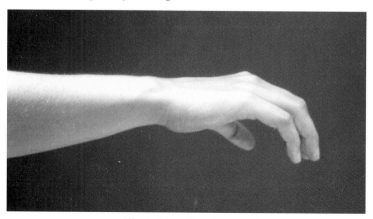

Illustrations Nos. 19 and 20

26

Do this until you feel your fingers relax and you feel some "give" in your knuckles. You'll notice that your fingers will become gently curved. Everyone's fingers curve in different ways. If you have long fingers, the curve will be a little less pronounced than it is in someone with short fingers.

You need to find the way your fingers naturally curve. To do this, drop a hand into your lap, or rest it on a table, with your fingertips down and completely relaxed. Look at your hand and notice the natural curve of your fingers, and then imitate it at your keyboard. Once again, you want your hands to look the same at your keyboard as they do everywhere else.

Whenever you notice your fingers getting into the Spider, gently raise up under the knuckles just a little until you can feel your fingers and your thumb relax.

Get used to spotting the Spider and fixing it instantly. You'll feel as if you're moving your hand from a collapsed position into one with a slight arch.

Think of a rainbow.

Chapter Seven

SORE THUMBS

The thumb can be so problematic that it needs its own chapter. If you have a sore thumb, you are probably typing in one of the ways shown below, all of which are actually recommended in various typing courses. The first is a classic typing position:

Bent Thumbs - No

This one is also popular:

Comma Thumbs - No

Another common and widely taught position is one in which you use your right thumb for the Space Bar while tucking your left thumb under your hand, like this:

Tucked-Under Thumb - No

All three of these positions can be trouble. They require you to hold your thumb tensed in an unnatural way, with its muscles continually contracted. When your thumb is tight, your whole hand is affected. Tense thumbs make for tight hands.

Try each of these "No" positions, one by one, away from your keyboard. Notice how each one immediately makes your hand and wrist tighten up. Touch the thumb muscles in your palm with the opposite hand and feel how rigid and hard they are after just a few *seconds* in one of these positions. Imagine holding your thumb like that *all day.*

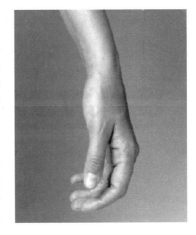

Every muscle in your body works by alternately contracting and relaxing. Your finger muscles are like every other muscle in your body — they need a chance to rest. If they don't get this chance — if they're required to *stay* contracted — they will quickly become tight and strained, and a tight muscle is easily injured.

The perfect thumb position is easy to find — it's one where your thumb is completely "at rest." Just drop your hand down at your side, relaxed, like this:

29

Keep that position at your keyboard, and it will look like this:

Illustration No. 25

Relaxed Thumbs - Yes

There may be a slight, natural bend at each knuckle, and there may not — people's thumbs are different. But there will be nothing extreme, as in Illustrations Nos. 21, 22 and 23. Your thumbs will feel completely free and loose.

Don't worry about where they fall on the Space Bar. Contrary to the advice of some typing courses, they don't need to be at some fixed location on it, such as "one-half the distance between B and N, (right) and B and V (left)", as one course recommended. The Space Bar goes down wherever you press it. Where *your* thumbs fall naturally is the right position for you. One of the advantages of avoiding the Dangerous Angle is that your thumbs never get in each other's way.

Should you use both thumbs to type? Yes, if you're used to it. It divides up the work. But, if you're used to using only your right thumb, and don't want to change, that's okay. Just make sure that you're not tucking your left thumb under your hand, or holding it up in the air. When you're not using your thumbs, rest them lightly on the Space Bar.

30

Use a small, easy, down-and-up motion — not sideways, or in-and-out. A straight down-and-up motion is less stressful than a sideways one.

You may notice that when your hand is relaxed, it will move slightly, and may actually rotate a bit with your thumb. That's fine — let it. One of the essential elements of safe typing is to *never resist natural movement*. If your hand wants to move with your fingers, don't stop it. The motion helps your fingers. When you're typing, this motion will be small and quick — almost imperceptible — but it's there.

EXERCISE No. 7
"Good Thumbs"

Drop your hand, relaxed, by your side.

Bring your hand up, still relaxed, and *look* at your thumb. This is its most natural position.

Now, at your keyboard, with your other fingers resting comfortably on the home keys, gently drop your thumb to depress the Space Bar and then release it. Use a small, down-and-up motion (not sideways).

Now try it quickly. Let your hand and forearm move slightly *with* the motion. Your thumb will move easily and comfortably and you'll feel no tension.

When you're not using your thumbs, rest them lightly on the Space Bar. Make sure you're not holding them in any of the "No" positions.

Learn good thumbs. They will contribute to the health of your whole hand.

 TIP: You'll find some good massage techniques to help keep your thumbs relaxed in Chapter 19.

Chapter Eight

RELAXING YOUR RINGS AND PINKIES

From Sore Thumbs we move to the other side of your hand to deal with your ring and pinky fingers. These can cause just as many problems as your thumb — your pinky because it's short, and your ring finger because it's the weakest of all your fingers. The key layout most of us use doesn't help, either.

It may surprise you to learn that the QWERTY keyboard (named after the first six characters on the third row) was designed in 1872, and *hasn't been changed since*. Its inventor thought that this particular pattern of keys would have two "advantages". The first was that you could spell TYPEWRITER without leaving the top row — meant to convince people that this forbidding, new machine might not be so difficult to learn after all. In fact, it was suggested that it might even be faster than a pencil.

But it couldn't be too fast. Because engineering and manufacturing were not advanced enough at that time to allow one key to get back to its place before the next one was pressed, the inventor, a Mr. Sholes, came up with an interesting solution: Make the pattern of keys so awkward and difficult that no one — not even an expert typist — could learn how to type quickly on it. That was the second "advantage" of the QWERTY keyboard.

Mr. Sholes believed that the last two fingers, the ring and pinky, were just *too weak to use*, and that everyone would use two — or at most four — fingers on his typewriter (two-finger typists, take heart!). This method, still popular today, became known as the "hunt-and-peck". I call it "running around your little finger" — named after tennis players who will run several extra steps to avoid using their backhand. I'm sure you've seen typists avoiding a particular weak finger by running around it all over the keyboard, thereby forcing all their other fingers into awkward positions.

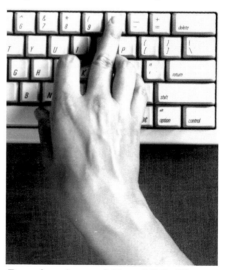

Running Around Your Little Finger

The unusual idea of using *all* fingers, known as "touch typing", became popular in 1888 when a speed contest was held in Cincinnati, pitting a "hunt-and-peck" typist against an "all-your-fingers". Not surprisingly, "all-your-fingers" won.

What's the relevance of this today? We are left with a key layout that was *purposely* designed to be difficult and awkward to use. There is an alternative, better layout called the Dvorak, but it requires learning a new system of typing and either buying a new keyboard or new software.

What makes the QWERTY design so difficult? The layout of the keys is such that your shorter and weaker fingers have to do the hardest work. Not only do they have to cover some of the most-used keys, they also have to reach for the awkward keys on the extreme right and left of the keyboard. If you reach for these keys incorrectly, you can seriously injure your hands.

Your pinky may be short, but it's strong. It has its own set of muscles, which you can feel on the underside of your hand. Your ring finger, however, is your weakest finger. Careless use of it alone is responsible for countless hand injuries, both to typists who use it continually and to those who "run around it ".

Why is the ring finger so easy to injure? This little experiment will quickly show you. Rest your right hand on a table, fingertips down. Now, *gently*, try to lift just your ring finger, leaving the other ones down. It's difficult. Your hand tightens up, and the finger moves only a fraction of an inch.

Now lift your middle and ring fingers together. It's much easier and your ring finger lifts twice as high. Try lifting your ring and pinky together. Also easy. All three at a time is the easiest. This is because the muscles that move your ring finger are completely bound up with those of your middle finger. Your ring finger can never be completely independent. *Injury to the ring finger is a result of trying to isolate it from your other fingers.*

If you're using either your ring finger or pinky incorrectly, it may feel sore, tired or tender. If it gets worse, the pain can radiate up your palm, through your wrist, and shoot all the way up your arm. You may also feel a "curling" sensation as though the finger wants to fold back into your hand. It may be uncomfortable just to lay your hand out flat.

These sensations can be the result of holding either or both of these fingers continually contracted. You want to avoid both flying your fingers up in the air ...

Flying Ring and Pinky - No

... and curling them under like this:

Curling Ring and Pinky - No

When either your ring finger or pinky is tight, it's virtually impossible to keep your hand relaxed. When these fingers are relaxed, your whole hand can relax:

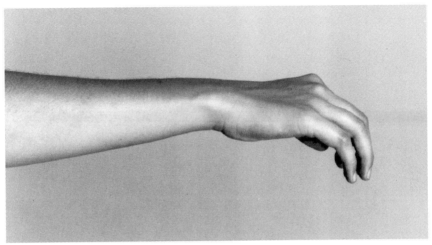

Illustration No. 29

At the keyboard it will look like this:

Yes

All of your fingers are part of an organic, curved whole. Place one hand over the other and feel the shape of this natural curve. Press down slightly, and feel how strong this curve is, how much weight it can easily support, and how each finger contributes to the support of the whole hand.

35

In the following exercise, you'll learn how to keep your ring finger and pinky always relaxed.

EXERCISE No. 8
"Relaxing Your Rings and Pinkies"

Drop your hands down at your sides, relaxed. Swing your arms a little. Wiggle all your fingers. Feel how your ring and pinky fingers move lightly and easily in this "dropped" position.

Now raise your hands to the keyboard and start typing gently. Keep that "dropped" feeling in your rings and pinkies — light and easy.

Make sure you're not flying your ring or pinky fingers, or curling them under.

If a finger starts feeling tight, or stubbornly persists in moving into a "No" position, drop your hand again. Feel the natural *shape* and curve of that finger when it's relaxed. See that natural shape in your mind. Try typing again, encouraging your finger to keep this shape at the keyboard.

If you have a problem finger, alternate between dropping and then typing, as many times as you need to until your finger can learn to keep its natural shape at the keyboard. Be patient.

Protect your rings and pinkies.

Your hands will thank you.

TIP: If you're a two- or four-finger typist, it can be painful and dangerous to hold certain fingers stiffly and awkwardly out of the way, just so you can see where you are on the keys (see Illustrations 26, 27 and 28 on pages 32 and 34). Even if you're not using your other fingers, keep them relaxed.

Chapter Nine

BUTTERFLY FINGERS

How can you keep *all* your fingers light and free?

The main thing to remember is that typing actually takes very little effort. I'm sure you've seen people who type by slamming their fingers into the keys, as if they're nailing them to the keyboard or destroying little bugs lurking underneath. Find the minimum amount of weight you need to push down a key. Remind yourself often how little weight that is, and catch yourself when you start "slamming." Think Butterfly Fingers — light, fast, fluttery, and free of tension. The little bit of weight you need to type will come from the relaxed weight of your arm, not from the "force" or "effort" of your fingers.

Many people think of their fingers as being almost like sticks — hard, inflexible bones, protruding out from their hands. Think of them as soft, curved and pliable. You don't need to *hold* them in a fixed position to type, any more than you need to freeze your wrists or glue your elbows.

When we think of control, speed and accuracy, we think we need to tighten up — as if that extra effort will produce those results. The opposite is true. Tightening up just makes typing (and everything else) much more difficult.

Perhaps the greatest compliment one can pay an athlete is to say that a particular movement he makes looks "effortless". A great athlete will actually look like he's doing almost nothing; it's as if the motion (think of a golfer) just passes through his body. Every part of his body remains relaxed and "boneless". You've certainly heard the remark, particularly before a crucial shot, "He's tightening up — he's going to blow it." Remember that. The tighter your fingers are, the slower and less accurate you'll be.

The best way to acquire light and fluttery fingers is to feel their opposite — tight, "stick" fingers. You'll dislike the stick feeling so much that you'll learn to stay relaxed without even having to think about it.

In the previous two chapters, we mentioned that tension and injury can be caused by certain extreme positions, but you may be surprised to learn that even holding up a finger a third of an inch off the keys can injure it.

Here's how to avoid "stick" fingers.

EXERCISE No. 9
"Butterfly Fingers"

Rest your hands on a table, fingertips down and completely relaxed.

Drum your fingers. Let them flutter freely and naturally.

Now gradually tighten them, making them harder and harder until they feel like they're turning into sticks.

Then, gradually, still drumming, let them go, feeling them become airy and light.

Do this again — tighten and untighten — and focus on how the muscles in your wrists and forearms feel when your fingers are tight. Untighten, and feel those muscles uncoil and relax.

Try to move your fingers really fast in the "stick" position — you can, but it takes a lot of effort.

Now do "Butterfly fingers" — flutter your fingers. They'll feel light, free, and relaxed.

When you're tense — maybe under a deadline — it's an easy thing for your fingers and hands to tighten up without your even realizing it. Just remind yourself of how little effort it takes to push down a key. Remember that you'll have greater speed and accuracy when your fingers aren't tight.

Think "Butterfly Fingers" — light and effortless. Whenever you need to remind yourself, do the exercise: Butterflies, sticks, then back to Butterflies.

Your fingers will prefer Butterflies.

Chapter Ten

RESTING AT "HOME"

In typing, "Home" refers to the middle row of keys: A, S, D, F, covered by your left pinky, ring, middle and index fingers respectively; and semi-colon, L, K and J, covered by your right pinky, ring, middle and index fingers. From "Home", your fingers travel out to cover all of the other keys.

One of the most crucial aspects of typing safely is learning a natural, comfortable, home position. Your fingers need to be able to rest at home — no pressing, no anchoring, no gluing — just resting. They should be able to maintain their home position for hours without feeling any strain. When your home position is uncomfortable, your fingers are in trouble before you've even started trying to reach the hard, outer keys.

The difficulties in maintaining that comfortable home position stem from a simple problem. When you hold your fingers straight out in front of you, you'll see that they are all different lengths. The keys, however, are in a straight line. How can you get your fingers "in line" with the keys without causing tension in your hand? Here's what you *don't* want to do — contort your fingers into what I call the Claw — the most widely taught home position:

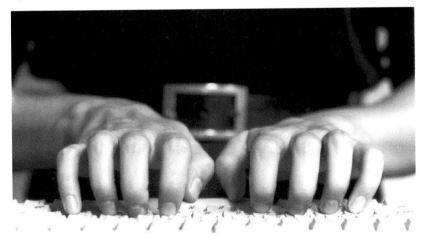

The Claw - No

The Claw is similar to the Spider, but instead of your hand knuckles being collapsed, they are flattened out into an even line. The best way to try out the Claw is to put your fingers out on a table, palm down, and then to draw them in as if you're scratching your finger-nails across a blackboard. Now hold for ten seconds. Try to pick up the Claw hand with your other hand. It will be glued with tension to the table.

Another way to flatten out your hand knuckles is with this contraption, invented in 1881 for the keyboard operators of that time (pianists) to work against nature and keep *their* knuckles in an even line:

Illustration No. 32

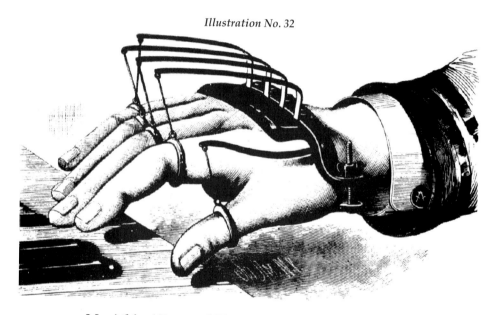

Mr. Atkins' Patented Finger-Supporting Device - No!

The right way to get your fingers in line with the keys? Don't work against nature. Rest your hand on a table, relaxed. Look at the line of your knuckles and you'll see a slight tilt down from your middle finger to your pinky.

This tilt of your knuckles is nature's way of compensating for the different lengths of your fingers — don't try to alter it. Keep the tilt at your keyboard. *Never resist natural positions.*

The Claw - No

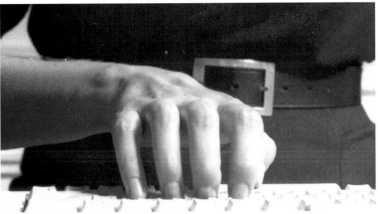

EXERCISE No. 10
"Resting at Home"

Rest your hand on a table, relaxed, and notice the tilt of your knuckles.

Go to your keyboard and put your fingers on the home keys. Keep your natural tilt.

Rest your thumbs naturally on the Space Bar.

Do a little "Butterfly Fingers".

Rest at Home.

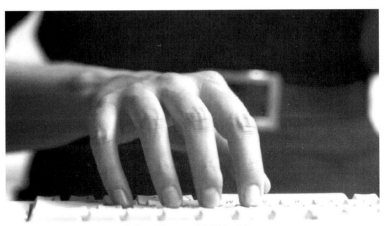

The Natural Tilt - Yes

Chapter Eleven

THE EVILS OF LIFTING AND SNAPPING

A typist's fingers travel about sixteen miles in an eight-hour workday. For the fingers, that's the equivalent of running a marathon *every day*. When you actually run a marathon, your stride is all-important. There can be no wasted motion, and you must be as energy-efficient as possible.

In Butterfly Fingers, we talked about avoiding unnecessary force while typing, and, in Resting At Home, about the most comfortable "home" position. What is the best "stride" for your fingers — the safest and most energy-efficient typing motion?

I came across some interesting answers to this question in various typing manuals. Some samples:

LIFT UP THOSE WEAK FINGERS!

PRACTICE YOUR HURDLES!

WORK THEM HARDER!

and ...

SNAP YOUR FINGERS!

USE A MOTION LIKE A SCRATCHING CHICKEN!

DOWN AND IN!

These ideas are interesting ... but unsafe. A little relevant anatomy will quickly explain why.

You lift your fingers up with the extensor muscles on the upper side of your forearm, which connect to tendons running down to your fingers. You press your fingers down with the flexor muscles on the underside of your forearm, also connected to tendons which run down to your fingers. When you're typing, your extensor muscles work *against* gravity, and your flexor muscles work *with* gravity.

Compare typing to walking for a moment. Walk around the room a little and notice how high you lift your legs when you take a step — just enough to clear the floor. Now lift your legs as if you were climbing stairs. There's a big difference.

Typing with high, raised fingers is like lifting your feet a foot off the ground each time you take a step. *Don't climb stairs with your fingers* — don't lift! Constant lifting is completely unnecessary and can seriously strain your extensor muscles. The keys are down, not up.

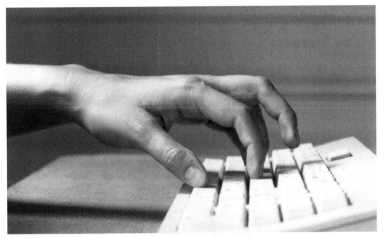

Climbing Stairs - No

When you type by snapping your fingers toward your palm, like a scratching chicken, you are also overcompensating, but in the opposite direction. *Don't scratch at the keys!* It's unnecessary and can strain your flexor muscles. The keys are right underneath your fingers, not somewhere back beneath your palm.

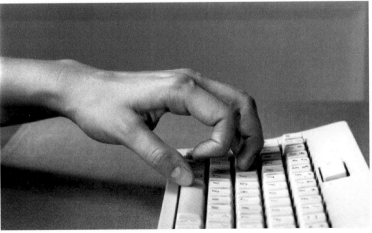

Scratching - No

The best motion is always the easiest, and in this case it's the one which works with gravity, not against it.

It's simple: *Drop* your fingers. Use a straight, easy, down-and-release motion:

Illustration No. 37

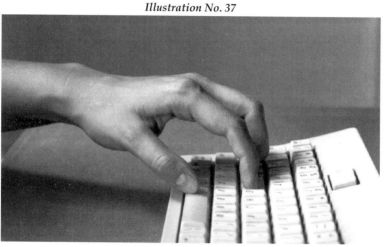

Dropping - Yes

The action of your fingers, the weight of your arm and gravity are now all going in the same direction.

Your finger muscles are not biceps — they're tiny and delicate. Nine tendons have to pass smoothly and freely through the small carpal tunnel in your wrist, as does the median nerve, which controls both sensation and movement in your hand. If you have the slightest swelling of any one of these tendons, the space inside the carpal tunnel can become constricted. Pressure is then put on the median nerve, and once that happens, you have the beginnings of serious hand problems. That is why it's so important to make sure you're using the easiest and least stressful finger-stride.

Dropping your fingers into the keys is one way to help lighten the stress on them. Here's another: One of the things which makes a computer keyboard so fast — the ease of pressing down a key — can also make it hard to use safely. Because the keys go down so easily, you may actually be holding your fingers slightly tensed over them, to prevent your fingers from pressing unwanted keys. Even this slight amount of tension is too much.

You'll want to learn "idling" — the simple coordination of keeping your unused fingers free and relaxed. Remember to float your wrists; it's essential for this.

EXERCISE No. 11
"Dropping and Idling"

Rest at Home. Feel your fingertips resting lightly on the keys.

Now drop and release both index fingers together on their respective keys several times. Feel how far down the key goes. Feel it spring back. Feel your fingers dropping and working with gravity.

Try it with your middle fingers, and then your ring and pinky fingers. Let your hands rock up slightly as the key goes down. As you release, and the key comes up, they'll naturally rock back.

Now do it quickly, using any fingers you want. Your hands will bounce lightly and your elbows will be loose and free.

Feel the interaction of your whole arm — feel how your arm weight can balance on your fingers. Feel your arms and gravity *helping* your fingers.

Concentrate on *idling* — when you're not using a finger, let it rest lightly on the keys.

Drop your fingers into the keys and idle the others.

Once you've found your stride, typing will feel great.

TIP: If you're a "snapper" or "lifter", you can easily break the habit. Do the exercise above several times, really getting used to the feeling of dropping your fingers and resting the others. Now alternate between your usual way and dropping. Your usual way will soon seem surprisingly uncomfortable. The choice won't be hard to make.

Chapter Twelve

THINK HANDS, NOT FINGERS

Once your fingers can easily rest at Home, and you've found your most comfortable finger-stride, how can you reach for the hard keys — the number row and the keys to the extreme right and left of the keyboard — without straining your hands?

This chapter is half of what you need to know. "Travelling Safely", the next chapter, is the other half.

Many people type as if ...

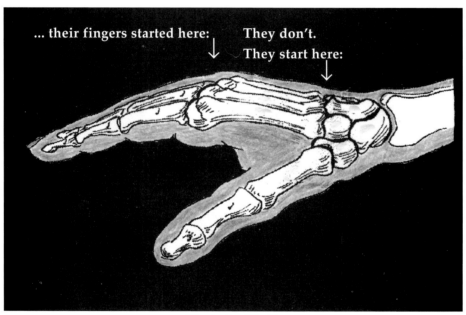

Illustration No. 38

Half of your fingers are inside of your hand. Your fingers are actually twice as long as what you see and normally think of as "fingers". When you reach, learn to use the whole finger — not just the half you can see.

The traditional way to reach for the hard keys is to anchor your fingers on Home and protrude a lone finger.

This is one of the most dangerous things you can do at your keyboard.

Reaching in this way collapses the natural arch of your hand and effectively cuts your finger in half at the bridge, breaking its natural, strong curve:

Protruding - No

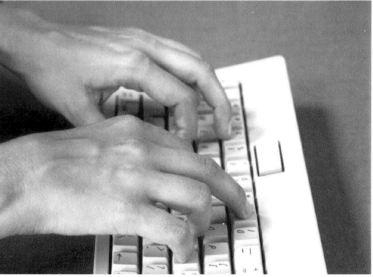

Preserving the Curve - Yes

Imagine holding a tennis ball and then tossing it up into the air. As you toss, your fingers spring open.

Imagine doing this a few times and get a sense of where the expanding motion begins — deep in your palm.

Now try the same motion on a table. Start with your fingers in a light fist, your hand relaxed. Then spring your fingers open:

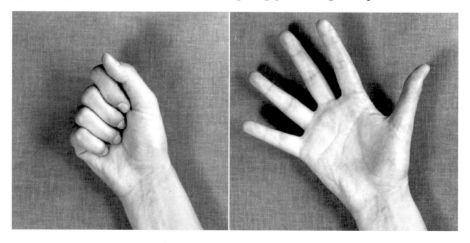

Once again, feel how the motion originates in your palm. Now try the same thing, but instead of the action being centered in your hand, center it in your fingers. It feels slower and less fluid, and your wrist tightens up.

When you reach for the hard keys, keep that natural, springy reflex of your fingers which you felt when you imagined tossing a ball. Expand from deep in your palm and lengthen your finger's natural curve.

Instead of anchoring your other fingers on Home while you're reaching, rest them lightly on Home. Then you can keep your whole hand, and particularly your bridge, loose and flexible.

Your fingers tend to mold and adjust themselves to whatever shape they're in contact with. Since your fingers naturally curve and your keyboard is flat this would seem to be a bit of a problem. But the following exercise will help your fingers keep their most natural and healthy shapes, regardless of which key you're typing.

EXERCISE No. 12
"Holding A Fruit"

Put your hand on a table, fingertips down, and imagine holding an orange. Feel the shape of the orange.

Expand your hand open, to the shape of a grapefruit. Your fingers won't reach all the way around.

Try these two shapes quickly, expanding from the orange to the grapefruit, then returning to the orange. Do it several times.

Go to your keyboard, rest your hands in the orange shape, and you'll see that it corresponds easily and comfortably to the home row. Slightly *expand* the curve of your fingers for the QWERTY row. Then bring it back to home. Slightly *condense* the curve for the low row, and then back to home.

Expand open to the grapefruit shape to reach the hard, outer keys.

Think of your fingers as a canopy draping lightly over the keyboard, every key easily accessible. When you reach, you'll expand from your palm, preserving the natural curve of your fingers.

Remember — don't protrude or curl.

Think hands, not fingers.

Chapter Thirteen

TRAVELLING SAFELY

In "Think Hands Not Fingers", you learned the safe hand-shapes which allow you easy access to all keys while preserving the natural curve of your fingers. There are also a few natural and simple motions for your wrists and arms which will add strength and momentum to your fingers, making typing even easier. You'll also increase your speed.

First, *forget* any rules you may have been taught such as:

DON'T BOUNCE!

KEEP YOUR HANDS PERFECTLY STILL!

KEEP YOUR WRISTS COMPLETELY FLAT AS YOU REACH!

On the contrary, *never resist natural movement* at your keyboard. The simple fact is that when your fingers and hands are relaxed, your wrists and arms will *want* to move in the direction of your fingers. Why prevent them?

Try a small experiment:

Put a small object, such as an eraser, on a table in front of you. Place your hand a couple of inches behind it. Now *reach* for it. Notice how you do it — the movement starts from the elbow, and your wrist rocks up slightly with the motion.

Now do the same thing and *prevent* any natural movement, keeping your wrist completely still and flat and not letting it "give" with the motion. It will feel stiff, awkward and slow after you've done it the natural way.

When you reach up to the number keys, which are a full *inch-and-a-half* away from Home, start the motion from the elbow and let your wrist "give". There may be a slight, natural, upward "rock" in your wrist.

When you reach down to the low row, again start the motion from the elbow, as if you're pulling out a drawer.

If you put the eraser a few inches up and to the right of your right hand and reach to touch it with your pinky, you'll notice how your hand slightly "rolls out" to the right as you reach. It's the same motion you'd use to turn a doorknob, only smaller.

Now try it again, preventing this natural roll-out, and keeping your wrist perfectly flat. You'll have to extend your pinky stiffly, and twist your wrist into the Dangerous Angle. Again, this will feel stiff and uncomfortable after you've done it the natural way.

Delete (or Backspace, depending on your keyboard) is almost *two-and-a-half* inches away from your pinky's home key, the semi-colon. Whenever you're reaching to the sides of your keyboard, let your wrists naturally roll, very slightly, *towards* the key.

It's simple: When you travel from Home, if you're keeping your wrists and hands perfectly still and extending lone fingers, you can easily strain and pull your finger muscles.

Travelling safely is easy: Let your hands and wrists move *with* your fingers. You don't need to think about it — just let your wrists follow your fingers.

Preventing movement is what takes the effort, and causes the damage. When you're typing fast, this movement will be slight — almost imperceptible — but your fingers will know the difference.

Will you be less accurate? No. Would a baseball player hit the ball more often, or harder, or farther, if he used just his arms to swing and concentrated carefully on making sure the rest of his body stayed perfectly still? Would a pitcher throw more accurately if he did the same? The idea is absurd.

Accuracy comes from the most efficient and *natural* use of the body.

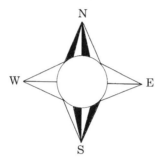

EXERCISE No. 13
"Travelling Safely"

Rest at Home.

Without typing — just concentrating on the motion — try travelling North, up to different number keys. Start the motion from your elbows and let your wrists "give" with the motion. They may rock up slightly with your fingers, and rock back down when you return to Home.

Now try going South, from Home to the low row. Again, start the motion with your elbows, as if you're pulling out a drawer, then return to Home.

Now try East and West, to keys such as Return and Delete (right hand), and Tab and Caps Lock (left hand). Think of the outside of your keyboard as an arc. Work your way around this arc, rolling out slightly with your hands and wrists, as if you're turning doorknobs, and then returning to Home. Your wrists will naturally adjust the motion for the various keys.

Now, actually type and try out these three motions. Keep the natural curve of your fingers, and let your hands and wrists move with your fingers. The three motions will blend into one fluid, continuous motion when you're typing fast.

Don't leave your hands and wrists behind when your fingers travel from Home. Never resist natural motion.

Travel safely.

TIP: As a last tip at the keyboard, here's an Instant Remedy for any bad position. If you see yourself getting into the Cobra, the Spider or any of the other harmful positions, drop your hand to your side and *keep* the bad position. Then slowly let your hand ease back into a relaxed shape. Feel that shape for five seconds — think and feel your way through your hand and fingers, and then go back to typing.

An important thing to remember is that one bad position encourages another, and if you're not careful you'll have something like an avalanche effect — first a sore thumb, then a tense pinky, then aching wrists. Doing the above remedy at the first sign of an incorrect position can prevent this from happening.

Chapter Fourteen

DON'T SQUEEZE THE MOUSE!

Your mouse may look innocent, but careless use of *it alone* can cause a serious hand injury, regardless of how safely you're using your keyboard. A friend of mine who was having some hand problems asked me this question: "When I use my mouse, my wrist is in the same position as it is when I do pushups. Is that good?" No. If your wrist looks like this when you're using your mouse ...

Illustration No. 45

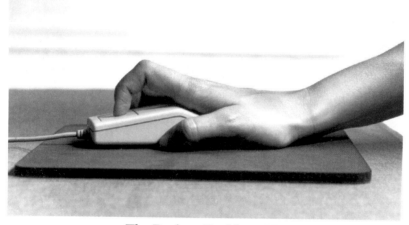

The Pushup Position - No

... you're back to the mousing equivalent of the Cobra position from Chapter Three.

Start by considering your mouse itself. Conventional mice are designed so that the clicker is somewhat higher than your wrist when it's resting on the working surface. Particularly if you're doing work which requires prolonged mouse use, such as desktop publishing, you'll want to make sure that you're not pressing your wrist down into the work surface while using your mouse. This forces your hand to bend up at the wrist, and, even worse, creates extreme pressure on the tendons and nerve which pass through the carpal tunnel.

As in safe keyboard use, the line of your forearm should be approximately level, with little or no bend in your wrist.

There are several new ergonomic mice available. In general they are somewhat larger than conventional mice, particularly at the back, and are designed to fit the shape of your hand. The clicker is at the front of the mouse, rather than on top, to ensure that your fingertips and wrist will be at approximately the same height. Also, some new mice have a click-lock function which enables you to drag without holding the clicker down, thus avoiding the strain of extensive pressing. As an alternative to the table mouse, you may want to consider using a foot-activated clicker. Used in conjunction with the arrow keys found on most keyboards, this allows you to insert your cursor without a trip to the mouse pad. There are mouse pads available which are built up at the back to support the wrist. If you use one, however, make sure that it allows free movement of your whole arm while you're using it.

Now for mousing techniques. Don't angle your wrist from side to side, cutting it off from your arm ...

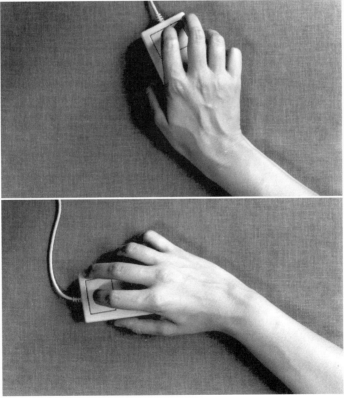

Angling Your Wrists - No

Use your whole arm, and keep the Natural Line (see page 9).

What is the proper way to hold a mouse? Don't strangle it ...

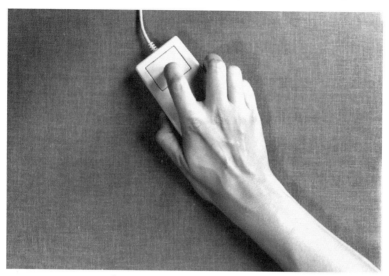

Squeezing the Mouse - No

... or point your finger at it:

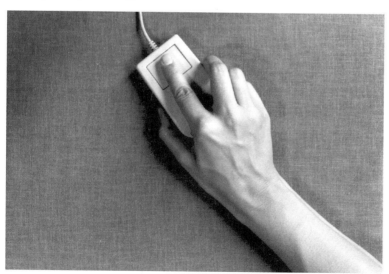

Pointing at the Mouse - No

Rest your hand on a table, relaxed. Now, keeping the same feeling, *drape* your fingers over the mouse. Your hand should feel completely relaxed and your fingers loose:

Illustration No. 50

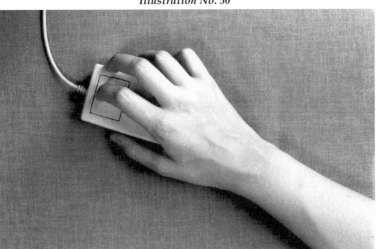

Draping the Mouse - Yes

Depending on the size of your fingers, you may want to try using two fingers (your index and middle fingers) to click your mouse. If this feels comfortable to you, it will only take a few days to get used to it, and this can considerably lighten the load on your index finger.

EXERCISE No. 14
"Mousing"

Drape your fingers over your mouse. Check to make sure you're not strangling it, or pointing your finger at it, or squeezing it with your thumb.

Now, with your computer off, draw circles with your mouse. First counter-clockwise, then clockwise. Start the motion from your shoulder and use *your whole arm.*

From circles, you can graduate to figure-eights.

Turn your computer on and practice using your mouse. Keep that fluid feeling, and you'll never feel the uncomfortable twist in your wrist.

TIP: If you're doing work which requires extensive use of a **numeric keypad**, follow the same principles as you would for safe mouse work. Make sure the keypad is positioned so that your wrist doesn't have to twist right or left, or bend up, when you're using it.

TIP 2: If you're using a **trackball** instead of a mouse, once again, follow the same principles. If you use your thumb, keep it in a relaxed, natural shape (you may want to refer back to Chapter Seven, "Sore Thumbs").

If you use your index finger, don't extend it stiffly, as in Illustration No. 49 (on page 56). Keep it relaxed, in its natural, curved shape.

Depending on the size of your fingers and the design of your particular keyboard, you may also find it comfortable to use two fingers on the trackball.

Part Two
Caring for Your Hands

Chapter Fifteen

LISTEN TO YOUR HANDS

A Repetitive Stress Injury does not happen overnight; no one just wakes up one morning with full-blown tendonitis or carpal tunnel syndrome. It is an injury which develops over days, weeks and months, which is why these injuries are also referred to as *Cumulative* Trauma Disorders.

An RSI will often begin with very mild symptoms — an annoying, slight cramp, a nagging, sore finger. If these mild symptoms are not immediately taken seriously, they can, over time, develop into a serious injury. Anyone who's suffered from a hand problem knows how alarmingly quickly a sore finger can become a sore hand, and how, within a few days, the pain can start radiating up the arm. You can stop this process before it ever gets started.

Listen to your hands. Never work through pain. If you feel any sensations of numbness, tingling, throbbing, cramping, or soreness while you're typing, stop. Take a fifteen-minute break and try some of the Hand Massage techniques in Chapter Nineteen, or the Four Stretches from Chapter Sixteen. Then try working again. If the sensation is still there, you may want to refer to the Twelve Golden Rules (APPENDIX B) to see if you're typing in one of the harmful positions. If you are, correct it. If the discomfort persists when you go home that night, you may want to use heat or ice therapy, whichever is most appropriate (see Chapter Twenty). If you still feel the problem after a few days, you may need to take some time off from your keyboard work, consult a physician, or both.

Your body has remarkable powers of recuperation if given a chance. Resting your hands and wrists — allowing the muscles some "downtime" — can give injured tissues a chance to heal. In some cases, a well-timed day or week off may be necessary in order to save months or years of crippling injuries. At the time, this may seem indulgent to both you and your employer, but it's not. Both the people who work at computers and their employers need to understand that therapy, a change of routine, and sometimes even a week off, can mean the difference between healthy hands and a case of carpal tunnel syndrome.

Pain and sensations of numbness, soreness, cramping, throbbing or tingling, even if slight, are signals to your body that a particular muscle or joint needs attention. Ignore the signal and you risk serious injury.

This does not mean that you should panic every time you feel a twinge. There is a difference between the momentary, passing twinge you can expect to feel occasionally with any hard-working muscle, and the more serious sensations mentioned above, which are the early warning signs of a Repetitive Stress Injury. A twinge will pass within seconds or minutes. A sensation which lingers may be the first indication of something more serious. Learn to tell the difference and take extra precautions when you need to.

You may feel a high degree of anxiety if you're experiencing even the slightest difficulties with your hands. In studies done on people with RSI, *all* sufferers reported a high level of psychological pressure.

If you need to consult a doctor, choose one who will discuss your symptoms in detail with you. If appropriate, your physician may suggest trying treatments such as deep-tissue massage and physical therapy before prescribing drugs which will relieve pain but may mask significant warning signals, and certainly before resorting to surgery. The decision on whether or not to have surgery is obviously a matter between you and the doctor. In some cases, it is the only option — for example, when electrodiagnostic tests reveal a clear-cut abnormality.

However, there have been no studies done on either the individual or long-term benefits of carpal tunnel surgery. In one study, done over a two-year period, of eighteen people who had carpal tunnel surgery, only one was able to return to the same job full-time with no recurrences of symptoms.

In itself, surgery is neither a cure nor a solution for sufferers of RSI. As stated in the Preface of this book, the Mayo Clinic has cautioned, "If followed by a return to the same traumatic environment, the operation is often unsuccessful in controlling symptoms." Surgery will only be successful when combined with retraining. This begs the question: If you're going to have to fix your keyboard habits *after* you've had surgery, why not fix them *before*? Maybe you won't need the operation after all.

Trust that you can learn how to deal responsibly and safely with your hands. Use Part One of this book, "At Your Keyboard", to learn how to fix any careless and harmful keyboard habits which may have been contributing to, or even causing, your injury. Use Part Two, "Caring For Your Hands", to learn techniques such as massage, heat and ice therapy, stretching and warmups to help you keep your hands healthy.

Chapter Sixteen

THE FOUR STRETCHES

Stretching your finger and arm muscles is one of the best ways to prevent injury and promote comfort, because it helps the muscles to maintain normal range of motion. When you use a muscle, you contract it, and when it's contracted, it's shorter. When a muscle is tight or strained, it stays stuck in this shortened state, unable to relax back to its original form. Stretching and lengthening out a muscle can help keep it healthy.

The following four stretches take only a few minutes, and if done daily can go a long way towards helping you stay injury-free.

You can use them as part of your Five-Minute Warmup (see the next chapter), and also while you're taking a break. However, if you have a pre-existing injury, check with your doctor before trying them.

The keys to doing these stretches properly:

- **Slow and gentle.** Stay completely relaxed during a stretch.

- Hold the stretches briefly — just a few seconds.

- Work within your limits. **Every stretch should feel good. If it hurts, stop.** If you feel any sensation of discomfort, no matter how slight, you're going too far. You should feel a pleasant pull, never pain.

- A stretch doesn't have to be extreme to help. Even a tiny stretch — as little as an eighth of an inch in some cases — can be beneficial. You'll notice that as your keyboard habits improve, your hands and fingers will become more relaxed and flexible.

- Stretches are best done when your hands are warm. Before starting, simply rub your hands together for a minute or so until they feel pleasantly warm.

Stretch No. 1
The Teepee
Stretches the Finger Flexors

In a teepee shape, with your fingertips touching, *gently* use the fingers of one hand to push away the fingers of the other. You'll feel a good stretch all along the inside of your fingers. Remember, a stretch doesn't have to be extreme — a little goes a long way.

You can do this with all your fingers at once, or one at a time. Don't forget your thumbs.

If you prefer, and if it's comfortable, you can also do this stretch on your desk or table by placing your fingertips on the edge and *gently* pressing down.

Illustrations Nos. 51 and 52

Stretch No. 2
The Curl
Stretches the Finger Extensors

Rest one hand on your table. Keeping it relaxed, gently push one finger at a time back towards your palm. Go only as far as is completely comfortable. You'll feel this stretch along the outside of your fingers and hand.

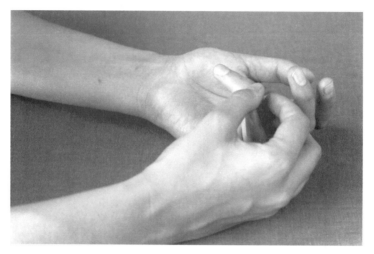

Then stretch your thumb by pressing it towards your palm:

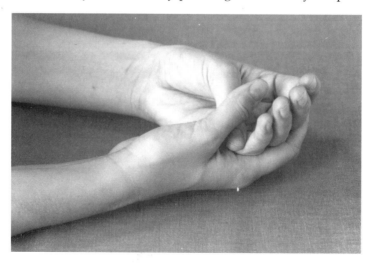

Stretch No. 3
The Press
Stretches the Forearm Flexors

Put your hands down in front of you, fingers of one hand resting on the palm of the other.

Lightly press down with your top hand while straightening your elbow — only as far as is comfortable.

You'll feel it along the inside of your wrist and forearm.

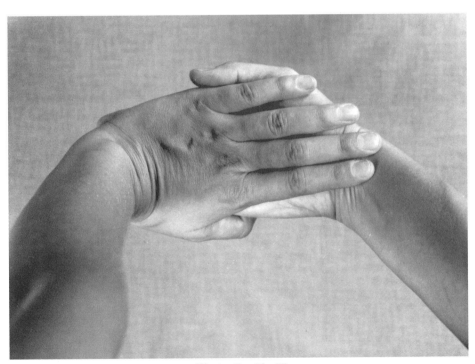

Illustration No. 55

Stretch No. 4
The Tuck
Stretches the Forearm Extensors

Keeping your hands dropped in front of you, cup the fingers of one hand in the opposite hand.

Gently, press down with your top hand while straightening your elbow.

You'll feel this stretch along the outside of your wrist and forearm.

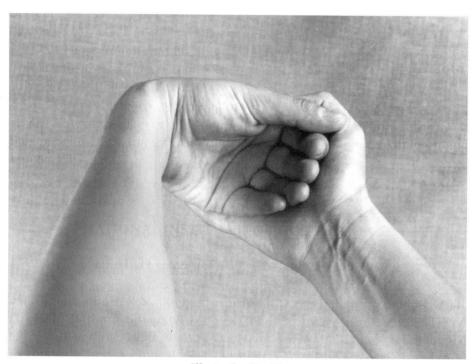

Illustration No. 56

After a week or so of doing these stretches, your hands and fingers will start having a wonderful, supple feeling.

Chapter Seventeen

THE FIVE-MINUTE WARMUP

Beware of starting to type with cold hands! Think of typing as a vigorous athletic activity for your fingers, and you'll naturally want to take a few minutes to warm up, just as you would if you were going for a run.

Your hands are valuable possessions — pamper them. Treat them carefully. If you've just had a week off, your fingers are going to be less flexible than usual. Allow a few extra minutes to warm up. Likewise, if you've just had a ten-hour day, the next day your fingers are going to feel creaky and tired. Again, take a few extra minutes and warm up gently, easing out the soreness.

The five-minute warmup consists of four easy parts:

1. The Rub

2. The Four Stretches

3. Finger Aerobics

4. Circles

If you have a pre-existing injury, check with your doctor before doing this warmup.

The Rub

Start with a good, light rub. Rub your hands together until you feel your palms and fingers heating up. Then rub the backs of your hands carefully and thoroughly, as if you're rubbing in lotion. Make sure you get all the nooks and crannies.

Roll up your sleeves (if you're wearing a long-sleeved shirt) and rub up and down the upper side of your forearm, all the way from your elbows to your fingertips and back. Linger on any sore spots. Then switch arms. Turn your arms over and do the inside of your forearm.

Lightly squeeze and massage the big muscle on the upper side of your forearm:

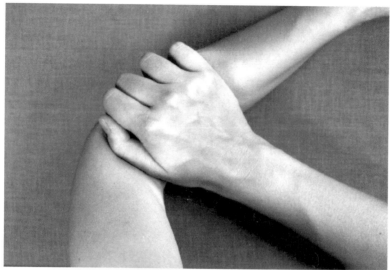

Illustration No. 57

Rest the palm of your hand against the underside of your forearm to warm it up.

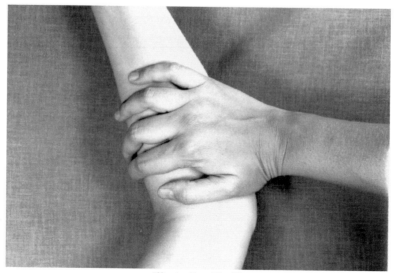

Illustration No. 58

Clasp your hands together and just rest for a few seconds. Enjoy the soothing heat from the light rub you just did. Feel the webbings between your fingers begin to warm up. Gently squeeze a few times.

The Four Stretches

Drop your hands down at your sides. If comfortable, lightly shake out your wrists. Make a fist and then spread out your fingers. Do this a few times. Then do the Four Stretches from Chapter Sixteen. All stretches should be done slowly and gently. If anything hurts, stop.

Finger Aerobics
Pushups and Pulldowns

Rest your hands on your desk, your fingers gently curved, with your fingertips and palms down. Raise your palms just off the surface and allow the gentle weight of your arms to settle onto your fingertips, the weight evenly distributed among all your fingers. Feel the strength of your natural arch supporting the weight. Apply a little more weight to your fingers and feel them beginning to warm up. You can move your wrists gently up and down and from side-to-side a few times.

Pushups

Now change the focus from all your fingers to just your thumbs. Slowly rock your wrists and arms up, letting your thumbs gradually take the weight.

Keep your other fingers completely relaxed. Hold just long enough to feel your thumbs warming up, and then slowly rock back down, taking the weight off. Feel your thumbs relax.

Go through your hands and do this on each individual finger. Rock up, apply a little arm weight, feel your finger warm up, and then rock back down, letting it relax. Do this slowly enough so that you can really feel the difference between the muscles warming up and then relaxing.

You'll want to go through your hands a few times until all your fingers feel pleasantly warm.

Pulldowns

For Pulldowns, you'll want to do the opposite: With your fingers still gently curved, fingertips down on your desk, start with your wrists in the "up" position. Gently pull *down,* once again letting your arm weight rest on your thumbs, your other fingers relaxed. As you rock back up, gradually take the weight off. Go through your hands one finger at a time, gently pulling down, and then relaxing up. Use as much arm weight as feels comfortable to you.

Circles

Finish your warmup with circles. Fingertips down, trace a few gentle circles with each finger, in both directions. Let your hand and wrist easily rotate with your fingers.

After you've gone through your fingers, leave your fingertips down, and do easy circles with your wrists, both ways. Let your elbows join in the circles, loosening up your arms.

Keeping your neck relaxed, do shoulder circles, first forward, then back. Do a few light circles with your neck.

Raise both shoulders all the way up and take a long, deep breath. Hold, and then, on a sudden exhale, drop your shoulders.

You're ready to go.

 TIP: All of the Hand Massage techniques in Chapter 19 are also excellent as warmups.

TIP 2: If you find that your hands and arms feel cold even after warming up, and after you've been typing, then the overall temperature of your office is too cold for you. Your arms and fingers should always feel pleasantly warm while you're typing. If they don't, you may want to try Armsocks™, which are designed to keep your forearms and wrists warm.

Chapter Eighteen

TAKING A BREAK

If your job requires you to do continuous keyboard work, you should rest your hands at least ten minutes every hour. Five minutes every half-hour is even better. Studies on RSI agree that this amount of rest is essential, and much more beneficial than a single mid-morning or mid-afternoon break.

Take every additional opportunity to rest your hands that you can — for example while reading over work you've just done. Pay special attention to resting your right hand, the one most susceptible to injury.

Constant sitting can be very hard on your body, particularly on your back and neck. Make sure that your head is evenly balanced on your shoulders. Try to move your head often while you're working and your neck will stay looser. Also, try to support yourself with your stomach muscles and your back muscles. Keep your spine long, with a small natural curve in your lower back.

Whenever you can, get up, walk around, yawn, stretch, fidget. Do all the things your body naturally wants to do. Get a long cord on your phone so you can take calls standing up.

Check with your doctor before doing these exercises if you have a pre-existing injury. Don't do anything which feels uncomfortable to you.

Hands and Wrists

As often as possible, even if you only have 30 seconds, just drop your hands down at your sides. Feel the blood flow into your fingers. Let your arms hang completely relaxed from your shoulders. Gently swing your arms back and forth, like a pendulum.

Do the Four Stretches at least once during the day, in addition to your warmup.

Elbows

Reach your arms straight out to the sides and give your elbows a good stretch. Fan your fingers out wide.

Illustration No. 59

Shoulders

Take a deep breath and raise your shoulders all the way up, hold, and then, on a sudden exhale, let them drop. Do this a few times, inhaling on the way up, and exhaling on the way down.

Now do one shoulder at a time. Raise your right shoulder, letting your left come down. Then reverse it.

Rest your hands on your legs and press your shoulders backwards, as if you're trying to make them meet behind your back. Relax. Now do the opposite. Try to make them meet in front of you. Relax again.

Do shoulder circles. Keeping your neck relaxed, rotate your shoulders, first forwards, and then backwards.

Arms

Clasp your hands and reach up. Take a deep breath and have a good stretch. Yawn. Make a light fist, and then stretch out your fingers. Try out these three clasps. They all feel good.

The Three Clasps

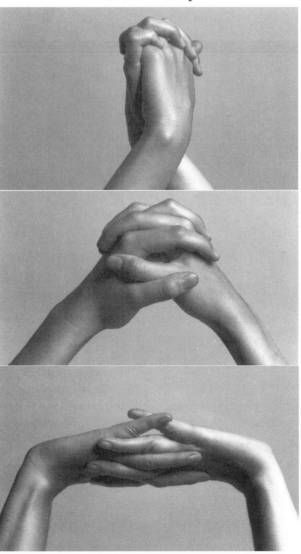

Tilt to each side, stretching under your arm and along your side.

Illustration No. 63

Clasp your hands behind your neck and push backwards with your elbows, stretching out your upper arm muscles.

Now, as if you're doing slow jumping-jacks, raise and lower your arms out to the side, making an arc. Inhale and fan out your fingers as your arms go up, exhale and let them relax as your arms come down.

Reach your hands in front of you, clasp your fingers, and stretch your arms slowly all the way up over your head. Look up at your hands, and then slowly relax back down. Do this a few times, inhaling on the way up, and exhaling on the way down.

Get up from your chair, cross your arms in front of you, and gently swing them back and forth to your sides.

Put your right hand on your left shoulder. Hold your right upper arm with your left hand and slowly bring it towards your chest.

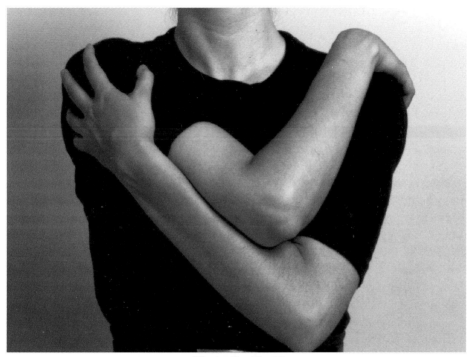

Illustration No. 64

Reverse sides.

Then put your right hand behind your neck, left hand on your right arm, and gently pull to the side.

Neck

Look straight ahead. Let your neck slowly drop forward towards your chest. Then raise it slowly, as if you're stacking your neck vertebrae on top of each other, one at a time.

Look up, following your eyes up to the ceiling, and then slowly come back down.

Look to your right, then back to center. Left, then back to center.

Tilt your neck from side to side.

Now do slow and luxurious neck circles, ironing out all the kinks. Lead with your chin and slowly roll your head forward, right, back, left and forward again. Do the opposite direction.

The Four Corners

Clasp your hands behind your neck. Press forward with your hands, and resist with your neck. Put your hands on your forehead, press, and again, resist with your neck. Now put your right hand on the right side of your head, press, and resist with your neck. Then do the left side.

Illustrations Nos. 65 and 66

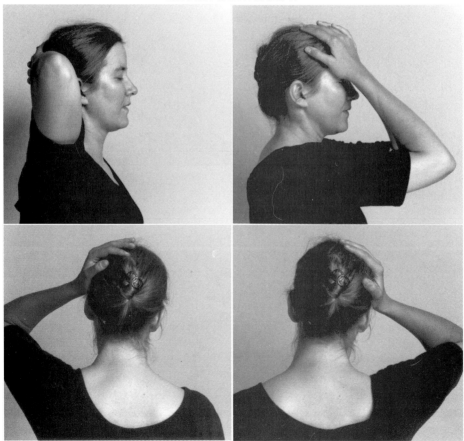

Illustrations Nos. 67 and 68

Finish off by giving yourself a little massage. First go up and down the back of your neck, using your fingers or your knuckles. Linger on any sore spots. Then massage both the right and left sides of your neck. Tilt from side to side and do a few more neck circles.

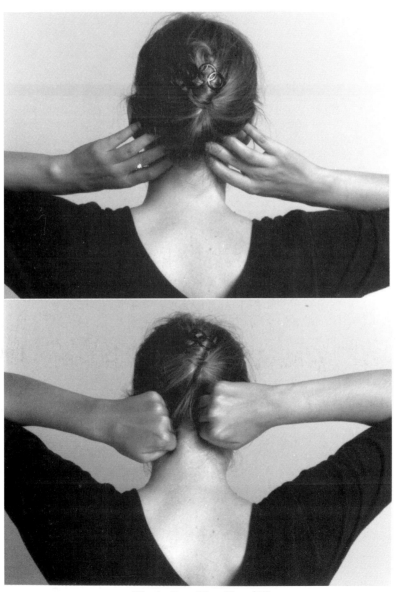

Illustrations Nos. 69 and 70

Back

Put your hands on your knees, inhale and arch your back. Look up. Stretch your shoulders back. Then exhale slowly, round your back, and let your shoulders drop. With one hand in front and one on your back, rotate very slowly toward your left.

Don't twist your back — first turn your head, and then continue rotating, using your *stomach* muscles to turn. Rotate as far as is comfortable, then slowly come back to center.

Reverse the hands and rotate in the opposite direction.

Stand up, drop your chin towards your chest, and slowly roll down towards the floor. Let your neck and arms hang. Slowly roll back up, using your stomach muscles.

Then stretch the opposite way. With your hands supporting your lower back, arch your back.

Now massage the ropey muscles on either side of your spine.

Chapter Nineteen

MASSAGING YOUR HANDS

When some part of your body hurts, you naturally want to rub it. It's almost a reflex. Learning how to massage your hands is invaluable in helping to prevent injury. It loosens stiff joints, eases the tension of tired muscles and improves flexibility and strength. The heat produced by a massage increases circulation to troubled areas and promotes healing.

There are two things you should be careful of while massaging your hands. First, how deep should you go? If you feel warmth, heat, even a feeling of pleasurable pain, that's fine. But if the "pleasurable pain" starts feeling suspiciously like just plain "pain", you've gone too far. Your hand will tense up against the pain, and you'll be creating an effect opposite to what you want. Hand massage should *always* feel good.

Second, pay close attention to the hand which is doing the massaging. If any of the pressing or rubbing techniques don't feel *completely* comfortable, stop. Instead, use something else. I've heard of suggestions ranging from golf balls to tennis balls, to walnuts, avocado pits and seashells. Be inventive — use what works for you.

 TIP: *Never* give yourself a massage with long nails. Nothing will feel worse to a sore finger than digging into it with a sharp nail!

Rest the hand that's being massaged on your lap or on any soft surface. You can also cradle it in your other hand. Keep the massaged hand completely relaxed. "Give" in to the pressure. Never resist — this will just make you tighten up.

You're going to come across some very sore spots while massaging, especially the first few times. Pay special attention to those areas. When they become too sensitive, move away to different fingers, and then return to those sore areas often during your massage. You'll soon get to know your hands and learn your trouble spots.

Again, if it starts feeling painful, don't fight it. Just lighten up. Go only as deep as is comfortable. A good massage hovers between pleasure and pain. Stay on the pleasure side.

Alternate frequently between your hands so that you don't tire out the hand that's doing the massaging. To start, rub your hands gently together for a minute or two to warm them up. Lay one hand down on your lap, palm up, and gently rub from the palm towards the fingers. Reverse the hands.

Turn them over and gently rub the outer side of your hand, from the wrist towards the fingers. Use your palm, fingertips, or thumb to rub — whatever feels most comfortable.

To warm up your fingers, take each one in your hand and give it a gentle hug.

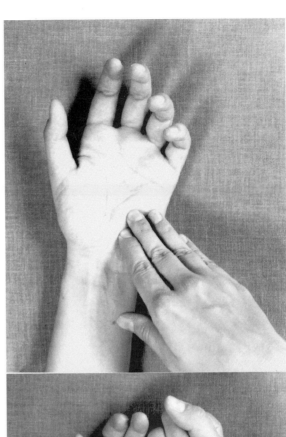

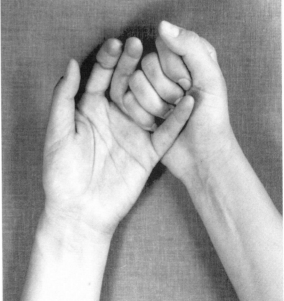

Illustrations Nos. 73 and 74

To massage a finger, start at the bottom and work your way slowly and gently up. Pay special attention to each knuckle and enjoy a good five seconds at the fingertip. Roll the fatty part between your fingers. Rub the sides, top and bottom — lightly work the whole finger. Alternate hands between each finger — right pinky, then left; right ring finger, then left. If you find an especially sore spot, rub it, massage some other fingers, and then come back to it.

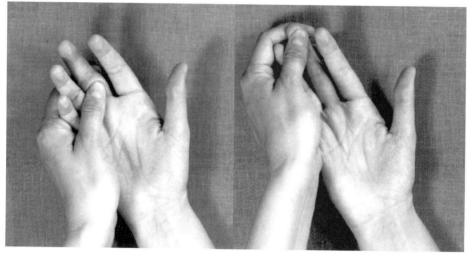

Now start to work on your hand — first the pinky muscle, then move up to the sensitive webbings between your fingers.

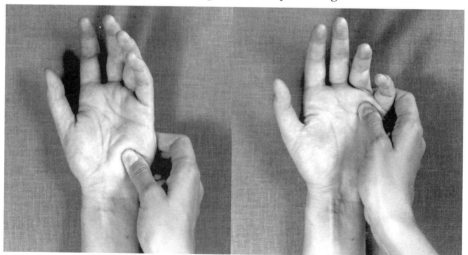

Illustrations Nos. 75, 76, 77 and 78

Remember to alternate hands. Go as deep as feels comfortable.

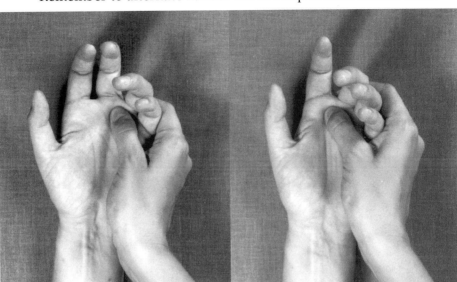

You may want to pay special attention to the webbing between your thumb and your index finger; this is often a very sore spot. Don't forget your large thumb muscle.

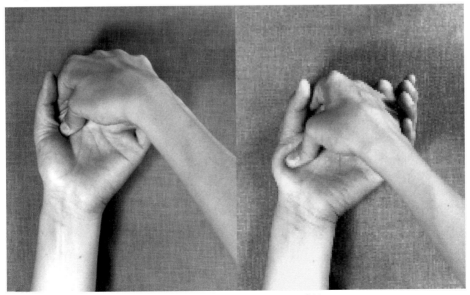

Illustrations Nos. 79, 80, 81 and 82

Turn your hands over, and massage your outer webbings.

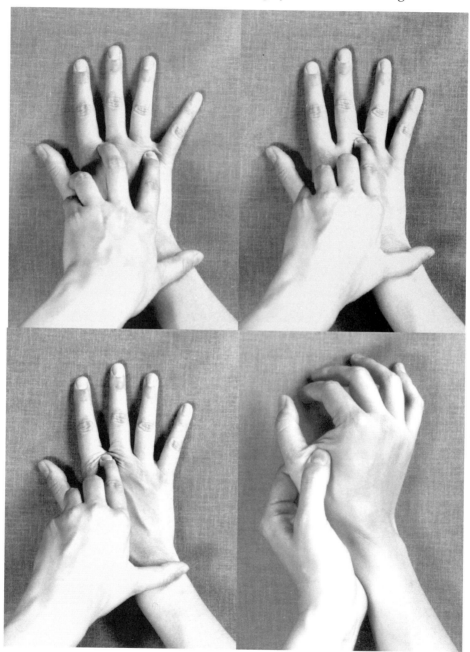

Illustrations Nos. 83, 84, 85 and 86

Now, turning your palms back up, trace the path of the whole finger — from the bottom of the palm, through the hand, and up into each finger-webbing.

After finishing your hands, lightly warm up both sides of your wrists. Move up your forearm and massage this point near your elbow:

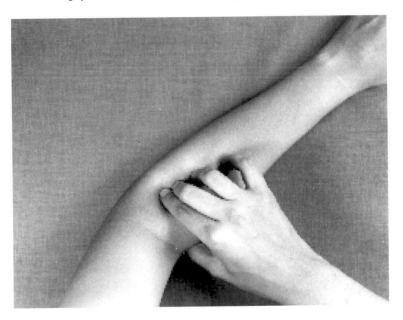

Then, as a little dessert, give your hands a light tickle.

It feels great.

Chapter Twenty

HOT AND COLD AND OTHER TIPS

Thermotherapy (heat treatment) and cryotherapy (cold treatment) can be highly effective ways to deal with injuries in their early stages and to help prevent them from becoming more serious.

COLD

Most doctors believe that ice is the best method of treatment during the first 48 hours of an injury. Ice reduces swelling and can provide immediate relief from pain. It can be applied in several ways. If you are experiencing problems with your hand, wrist or elbow, you can try applying an ice pack several times a day. If it feels as though it's burning your skin, wrap it in a towel or dishcloth.

If you don't have an ice pack, a bag of frozen peas will work just as well. You can also fill a paper cup with water, freeze it, and then apply the ice-cup to the sore area.

If you're experiencing finger problems, it can be a bit awkward to apply an ice-pack. A better way is to fill a pitcher with ice-water and then immerse your hand in the water for ten or fifteen minutes. If you find that too painful, as many people do, try this: Fill a pitcher with lukewarm water, add two trays of ice cubes, and immerse your hand quickly, before the water becomes too cold. This gives your hand time to adjust as the water gradually becomes more icy.

HOT

When all signs of swelling are gone, heat is very effective in increasing circulation, soothing tired muscles and helping them relax. Use a hot washcloth or heating pad, soak your hands and arms in hot water, or try using a hydrocollator (heat pack).

The alternation of hot and cold can also be very beneficial after any swelling has gone down. You can alternate ten or fifteen minutes of immersion in hot water with immersion in ice water. You can also alternately apply a heating pad and an ice pack.

OTHER TIPS

Don't type with long nails! They will force your fingers into an unnatural, stiff and flattened position at the keyboard.

When you are carrying heavy bags, try to hold them in your arms rather than carrying them down at your sides, which will force your fingers to bear all the weight. Even better, whenever possible, use a backpack for heavy items.

When you're lifting a heavy pan in the kitchen, try to use two hands rather than one.

While writing, hold your pen or pencil as loosely as possible to avoid strain.

Adjusting Your Workstation

APPENDIX A

ADJUST YOUR EQUIPMENT - NOT YOUR BODY

For optimal working conditions, three different heights must be adjustable at your workstation — the height of your chair, the height of your keyboard, and the height and tilt of your screen. When these three heights have been correctly adjusted, you will have created three approximately straight planes, which will allow your body to maintain its most comfortable and natural position, as shown on the opposite page. Use this illustration as a guide, but remember that everyone's body is different, and you'll want to adjust your workstation to fit *your* body.

The line of your thighs should either be level and parallel to the floor, or your knees should be *slightly* lower than your hips. The line of your forearms should be approximately level and parallel to the floor, with little or no bend in your wrists. And third, your line of sight to the screen will be either straight, or variable downwards from ten to 20 degrees.

1. Adjusting your chair height

The first adjustment to make in optimizing your working conditions is chair height. This is done in three easy steps:

First, sit on your chair with your feet flat on the floor. Take a yardstick or tape measure and measure the distance from the floor to the underside of your thigh, just behind your knee (measurement No. 1).

The second step is to find the correct height of your seat pan. Start by measuring the distance from the floor to the front edge of your chair. Then, if your chair has a soft seat, account for this by pressing down on it with one hand, and with the other hand measure how many inches it "gives". Subtract the "give" from the front-edge height, and you have measurement No. 2.

Thirdly, adjust your chair so that measurement No. 2 (the height of the seat pan) is equal to measurement No. 1 (the distance from the floor to the back of your knee). At this height, your thighs will be at an approximately 90-degree angle to your body and calves.

Try out your new seat-height. It should feel comfortable. The line of your thighs should be level and parallel to the floor. The weight should feel evenly balanced between your buttocks and along your thighs. If you feel any pressure-points, you may need a little further adjustment. If your feet do not rest comfortably flat on the floor, you may want to use a footrest. Finally, if the back of your chair does not provide enough support for your lower back, try using a small pillow to support your lumbar region.

2. Adjusting your keyboard height and tilt

In the prevention of hand injuries, particularly carpal tunnel syndrome, the most important height to adjust is that of your keyboard. When you're typing, the line of your forearms should be approximately level and parallel to the floor, with little or no bend in your wrists.

The easiest way to find this correct keyboard height? Sit in your chair at your newly-adjusted chair height with your forearm at a right angle to your upper arm. Measure from the floor to the underside of your elbow. Your keyboard, and specifically the home-row, should be at or slightly below this height.

Adjusting the tilt of your keyboard is a very individual matter. As mentioned on page 19, the device which enables you to increase the height of the back of your keyboard, so that the keyboard tilts *towards* you, with the space bar at a lower level than the top row, must be used cautiously. Too extreme a tilt will create the Cobra position. Many people find that they prefer no tilt at all.

3. Adjusting your screen height and tilt

The top of your screen should be at about eye level or slightly lower. Sit approximately two feet from your screen, nod your head up and down a few times, and find exactly the line of sight where your neck feels the least strain. Adjust the height of your monitor accordingly. Tilting the screen slightly may help you to maintain this comfortable, neutral neck position. When optimally adjusted, the tilt of the screen will be perpendicular to your line of sight.

4. Screen lighting and glare

You should have no bright light, either direct or reflected, in your line of vision, including the periphery. Adjust the room light —

both incoming light from windows and any artificial lighting — so that their relative brightness is approximately proportional to the brightness of your screen. A foreground which is considerably brighter than your screen may contribute to eyestrain. Likewise, sitting in a dark room and staring at a bright computer screen for hours can cause your eyes to become fatigued.

Harsh overhead light can create a harmful glare on your screen. If you can, adjust your overhead lighting so that it doesn't create glare when you're looking at your monitor. If you can't adjust the overhead lighting, you may want to buy an antiglare filter which fits onto the front of your screen.

Finally, keep your screen clean, and don't forget that, if you wear glasses or contacts, your prescription should be optimized for the usual distance from your eyes to your screen.

5. Air quality and temperature

Try to keep your work-area as well-ventilated as possible. Lack of sufficient air-flow can inhibit the dispersal of air-borne pollutants, and this may not only contribute to respiratory problems but may irritate your eyes as well. The temperature of your office should be such that your arms and hands feel pleasantly warm while you're working. As stated in Chapter Eighteen, working with cold hands and fingers can be harmful. If your office temperature is too low, and is not adjustable, wear either a sweater or Armsocks™.

6. If your equipment is not adjustable ...

The height of your chair may not be adjustable. Until you can get a chair which can be adjusted, don't compensate by sloping your fore-arms either up or down to meet the keyboard, or by bending your wrists into the Cobra or any of the other dangerous positions. Instead, as a *temporary* solution, try altering the chair height with pillows or a foam cushion. Another solution is to mount your chair on larger casters.

You can raise the height of your monitor by placing a telephone book, or the equivalent, underneath it. Then you can avoid having to tilt your head too far down to see the screen, which places strain on your neck muscles.

Finally, if you're using a notebook computer, first adjust the height of the keyboard as described above, and then adjust the tilt of the screen.

Appendix B

SUMMARY - THE 12 GOLDEN RULES

| Avoid the Dangerous Angle | Keep the Natural Line |

1

| Avoid the Cobra | Float your wrists |

2

| Don't glue your elbows | Let them hang free |

3

4

| Don't slam into the keys | Think Butterfly Fingers |

Avoid the Spider

Relax into the Rainbow

5

Avoid the Claw

Let your knuckles tilt

6

Don't exhaust your thumb

Have a good thumb

7

Don't fly or curl your rings or pinkies   Relax them

8

Don't lift or snap your fingers   Drop them

9

Don't protrude   Preserve the curve

10

11

| When you're travelling from Home, don't resist natural movement | Travel safely — Let your hands and arms move with your fingers |

12

When you're not travelling,
Rest at Home

And don't forget:

Don't squeeze the mouse!

The HAND Book

Bibliography

American Society for the Surgery of Hands, *Hand - Primary Care of Common Problems.* Churchill, 1990.

James Bennett and S. J. Wanous, *Professional and Personal Keyboarding and Typewriting.* Cincinnati, Ohio: South-Western Publishing Co., 1988.

Devaki Berkson, *The Foot Book.* New York: HarperCollins, 1992.

Judith Chiri, Jacquelyn Kutsko, Patricia Seraydarian and Ted Stoddard, *Keyboarding.* Boston: Houghton Mifflin, 1987.

Gaylord Clark, *Hand Rehabilitation.* Churchill.

Cromwell and Lehmann, *Hand Rehabilitation in Occupational Therapy.* Haworth Press, 1988.

Tammy Crouch and Michael Madden, *Carpal Tunnel Syndrome and Overuse Injuries.* Berkeley, California: North Atlantic Books, 1992.

Betty Etier and A. Faborn Etier, *Individualized Typing.* Indianapolis: Bobbs-Merrill Educational Publishing, 1983.

William J. Faber, D. O. and Morton Walker, D. P. M., *Pain, Pain Go Away.* San Jose, California: Ishi Press International, 1990.

Paul Davidson, M.D., *Chronic Muscle Pain Syndrome.* New York: Random House, Berkley Books, 1989.

Charles Duncan, Susie Van Huss, S. Warner, *College Keyboarding and Typewriting.* Cincinnati, Ohio: South-Western Publishing Co., 1991.

Michael Reed Gach, *Arthritis Relief At Your Fingertips.* New York: Warner Books, 1990.

David Harding, Kenneth Brandt, M.D., and Ben Hillberry, Ph. D., *Minimization of Finger Joint Forces and Tendon Tensions in Pianists.* Reprinted from MEDICAL PROBLEMS OF PERFORMING ARTISTS. Philadelphia, PA: Hanley and Belfus, Inc. 1989.

Ronald Harwin, M.D., and Colin Haynes, *Healthy Computing.* New York: American Management Association, 1992.

Stephen Hochschuler, M.D., *Back In Shape.* Boston: Houghton Mifflin Company, 1991.

Matthew Hoffman and William Le Gro, *Disease Free.* Emmaus, Pennsylvania: Rodale Press, distributed by St. Martin's Press, 1993.

The Johns Hopkins Medical Handbook. Random House, 1992.

Thomas Langford, *Basic Keyboarding/Typewriting Drills.* Cincinnati, Ohio: South-Western Publishing Co., 1987.

Kate Lorig, R. N., Dr. P. H., and James F. Fries, M. D., *The Arthritis Helpbook.* Reading, Massachusetts: Addison-Wesley Publishing Company, 1990.

K. A. Mach, James La Barre and William Mitchell, *Keyboarding and Applications.* Paradigm Publishing, 1990.

Laura Norman, *Feet First.* New York: Simon and Schuster, 1988.

Scot Ober, Robert Poland, Robert Hanson, Albert Rossetti, Alan Lloyd, Fred Winger, *Gregg College Typing.* New York: McGraw-Hill, 1989.

Kathryn G. Parker and Harold R. Imbus, *Cumulative Trauma Disorders.* Chelsea, Michigan: Lewis Publishers, 1992.

David Pisetsky, M.D., Ph.D. and Susan Flamholtz Trien, *The Duke University Medical Center Book of Arthritis.* Ballantine Books, 1992.

Cortez Peters, *Championship Keyboarding, Skillbuilding and Applications.* New York: McGraw-Hill, Inc., 1989.

Vern Putz-Anderson, *Cumulative Trauma Disorders.* National Institute for Occupational Safety and Health, Cincinnati, Ohio: Taylor and Francis, 1988.

Jerry Robinson, Lee Beaumont, T. James Crawford, Lawrence Erickson and Arnola Ownby, *Basic Information Keyboarding Skills.* Cincinnati, Ohio: South-Western Publishing Co., 1988.

Steven Sauter, *Improving VDT Work.* University of Wisconsin, Department of Preventive Medicine, 1985.

Dava Sobel and Arthur C. Klein, *Arthritis: What Works.* New York: St. Martin's Press, 1989.

Robert Spinner, M.D., John Bachman, M.D., Peter Amadio, M.D., *The Many Faces of Carpal Tunnel Syndrome.* The Mayo Clinic Proceedings, July 1989, Volume 64.

Lawrence Tierney, Stephen McPhee, Maxine Papadakis, and Steven Schroeder, *Current Medical Diagnosis and Treatment.* Appleton and Lange, 1993.

NATIONAL INSTITUTE FOR OCCUPATIONAL SAFETY AND HEALTH
HEALTH HAZARD EVALUATION REPORTS:

Susan Burt, R.N., Richard Hornung, Dr. P.H., Lawrence Fine, M.D., Dr. P.H., Barbara Silverstein, Ph.D., Thomas Armstrong, Ph.D., *HETA 89-250-2046 June 1990 NEWSDAY, INC., Melville, New York*, Hazard Evaluations and Technical Assistance Branch, Division of Surveillance, Hazard Evaluations and Field Studies, NIOSH, Cincinnati, Ohio.

Bruce Bernard, M.D., MPH, Steven Sauter, Ph.D., Martin Petersen, Ph. D., Lawrence Fine, M.D., Ph.D., Thomas Hales, M.D., *HETA 90-013-2277, January 1993, LOS ANGELES TIMES, Los Angeles, California.* Health Hazard Evaluation Report, NIOSH.

Thomas Hales, M.D., Steven Sauter, Ph.D., Marty Petersen, Ph.D., Vern Putz-Anderson, Ph.D., Lawrence Fine, M.D., Dr. PH, Troy Ochs, M.S., Larry Schleifer, Ed.D., Bruce Bernard, M.D., M.Ph., *HETA 89-299-2230, July 1992, US WEST COMMUNICATIONS, Phoenix, Arizona; Minneapolis, Minnesota; Denver, Colorado.* Health Hazard Evaluation Report, NIOSH.

Teresa M. Schnorr, Ph.D., Michael J. Thun, M.D., M.S., William E. Halperin, M.D., M.P.H., *HETA 85-452-1698 May 1986 AT&T, SOUTHERN BELL AND UNITED TELEPHONE, North Carolina.* Health Hazard Evaluation Report, NIOSH.

U.S. DEPARTMENT OF LABOR, BUREAU OF LABOR STATISTICS
ANNUAL SURVEY OF
OCCUPATIONAL INJURIES AND ILLNESSES FOR 1991
November, 1992, Washington, D.C.

SOFTWARE

Mavis Beacon Teaches Typing. Novato, California: The Software Toolworks, Inc., 1991.

Dvorak on Typing. Santa Ana, California: Interplay Productions and Park Place Production Team, 1991.

DEDICATION

To my husband, my chief supporter and critic, who was by my side through every step of this book and without whom it could not have been written.

ACKNOWLEDGMENTS

I would like to thank my family — my mother, Sharon Jones, without whom I would not have become a pianist, my sister, Bonnie, for her never-ending optimism, support and help to me while I was writing this book, my brother, Randy, for his invaluable counsel and advice, and my sister, Natalie, for her enthusiasm.

I would also like to thank Ruth Gamble, who showed me how to massage my hands, and Henry Shapiro, for passing on to me some of the concepts found in Chapter Twelve.

Photography
by
Terry deRoy Gruber

Appendix Illustration
by
Lydia Rivera

Stephanie Brown is a concert pianist who has appeared at Carnegie Hall, Lincoln Center, the Kennedy Center and the White House. She has performed with such American orchestras as the New York Philharmonic and the San Francisco, Detroit and St. Louis Symphonies, as well as with orchestras in Europe and Japan.

She began teaching in 1982 and has given lectures and demonstrations at over 100 colleges in the United States. In the course of this work, she encountered many pianists with hand injuries, some so severe that they were considering changing professions. While working with them, she began to see a direct correlation between harmful finger and hand positions and motions used at the keyboard and subsequent injury. Over the next several years Ms. Brown developed a method to teach pianists how to relax and feel comfortable at the keyboard — how to play without pain.

After using typewriters for twenty years, she bought a computer, and when she hired teachers to help her learn how to operate it, she quickly noticed that many of them used exactly the same harmful positions and motions she had observed in pianists. When asked if their hands hurt, they gave answers such as: "Yes, they ache at night and wake me up"; or, "My thumbs are always sore and tender"; and, "I have trouble cutting vegetables".

She realized that there was a real need for a book showing computer users how to type safely, and who better qualified to write it than a pianist — someone who has spent her life at keyboards.

"The HAND Book" is Stephanie Brown's first book.